Handbook of Micrometeorology

ATMOSPHERIC AND OCEANOGRAPHIC SCIENCES LIBRARY

VOLUME 29

Editors

Lawrence A. Mysak, *Department of Atmospheric and Oceanographic Sciences, McGill University, Montreal, Canada*
Kevin Hamilton, *International Pacific Research Center, University of Hawaii, Honolulu, HI, U.S.A.*

Editorial Advisory Board

L. Bengtsson	Max-Planck-Institut für Meteorologie, Hamburg, Germany
A. Berger	Université Catholique, Louvain, Belgium
P.J. Crutzen	Max-Planck-Institut für Chemie, Mainz, Germany
J.R. Garratt	CSIRO, Aspendale, Victoria, Australia
G. Geernaert	DMU-FOLU, Roskilde, Denmark
M. Hantel	Universität Wien, Austria
A. Hollingsworth	European Centre for Medium Range Weather Forecasts, Reading, UK
H. Kelder	KNMI (Royal Netherlands Meteorological Institute), De Bilt, The Netherlands
T.N. Krishnamurti	The Florida State University, Tallahassee, FL, U.S.A.
P. Lemke	Alfred-Wegener-Institute for Polar and Marine Research, Bremerhaven, Germany
P. Malanotte-Rizzoli	MIT, Cambridge, MA, U.S.A.
S.G.H. Philander	Princeton University, NJ, U.S.A.
D. Randall	Colorado State University, Fort Collins, CO, U.S.A.
J.-L. Redelsperger	METEO-FRANCE, Centre National de Recherches Météorologiques, Toulouse, France
R.D. Rosen	AER, Inc., Lexington, MA, U.S.A.
S.H. Schneider	Stanford University, CA, U.S.A.
F. Schott	Universität Kiel, Kiel, Germany
G.E. Swaters	University of Alberta, Edmonton, Canada
J.C. Wyngaard	Pennsylvania State University, University Park, PA, U.S.A.

Handbook of Micrometeorology

A Guide for Surface Flux Measurement and Analysis

Edited by

Xuhui Lee

*Yale University,
New Haven, CT, U.S.A.*

William Massman

*United States Department of Agriculture/Forest Service,
Fort Collins, CO, U.S.A.*

and

Beverly Law

*Oregon State University,
Corvallis, OR, U.S.A.*

KLUWER ACADEMIC PUBLISHERS
DORDRECHT / BOSTON / LONDON

A C.I.P. Catalogue record for this book is available from the Library of Congress.

ISBN 1-4020-2264-6 (HB)
ISBN 1-4020-2265-4 (e-book)

Published by Kluwer Academic Publishers,
P.O. Box 17, 3300 AA Dordrecht, The Netherlands.

Sold and distributed in North, Central and South America
by Kluwer Academic Publishers,
101 Philip Drive, Norwell, MA 02061, U.S.A.

In all other countries, sold and distributed
by Kluwer Academic Publishers,
P.O. Box 322, 3300 AH Dordrecht, The Netherlands.

Printed on acid-free paper

All Rights Reserved
© 2004 Kluwer Academic Publishers and copyright holders as indicated on appropriate pages within.
No part of this work may be reproduced, stored in a retrieval system, or transmitted
in any form or by any means, electronic, mechanical, photocopying, microfilming, recording
or otherwise, without written permission from the Publisher, with the exception
of any material supplied specifically for the purpose of being entered
and executed on a computer system, for exclusive use by the purchaser of the work.

Printed in the Netherlands.

Contents

Contributing Authors		ix
Preface		xiii

1 Introduction   1
Beverly Law, Shashi Verma

2 Averaging, Detrending, and Filtering of Eddy Covariance Time Series   7
John Moncrieff, Robert Clement, John Finnigan, Tilden Meyers
- 1   Introduction   7
- 2   Averaging, Detrending, and Filtering Operations   11
  - 2.1   Time averaging   11
  - 2.2   Linear detrending   12
  - 2.3   Filtering   14
  - 2.4   Advantages and disadvantages of the three methods: Their effect on the spectra   17
- 3   Choosing an Averaging Period or Filter Time Constant   18
- 4   The Origin of Low Frequency Content in the Signal   22
- 5   Errors Associated with Averaging, Filtering and Detrending   25
- 6   Averaging and Filtering in Complex Terrain: Special Considerations   27
- 7   Conclusions   29
- 8   References   30

3 Coordinate Systems and Flux Bias Error   33
Xuhui Lee, John Finnigan, Kyaw Tha Paw U
- 1   Introduction   33
- 2   Theory   35
  - 2.1   Mass balance at a point   35
  - 2.2   Coordinate systems   39
  - 2.3   Advantages and disadvantages of Cartesian and streamline coordinate systems   43
- 3   Coordinate Systems for Point Measurements   44
  - 3.1   General considerations   44
  - 3.2   Instrument coordinate   45
  - 3.3   Natural wind coordinate   47

		3.4	Planar fit coordinate	48
	4		Flux Bias Error due to Coordinate Tilt	50
		4.1	Momentum flux bias	51
		4.2	Scalar flux bias	51
	5		Examples of Coordinate Tilt	53
	6		Analysis of a Sample Dataset	55
		6.1	Dataset	55
		6.2	Results	56
	7		Conclusions	58
	8		Appendix A: The Natural Wind Coordinate System	60
	9		Appendix B: An Alternative Method for Rotation into the Planar Fit Coordinate	61
	10		Acknowledgment	64
	11		References	64

4 Uncertainty in Eddy Covariance Flux Estimates Resulting from Spectral Attenuation   67

William Massman, Robert Clement

1	General Issues Regarding Flux Attenuation		67
2	Sources of Uncertainties for the Transfer Function Method		73
	2.1	A general mathematical expression for spectra and cospectra	73
	2.2	Analytical expression for estimating uncertainty	75
3	Application of Uncertainty Analysis to Observed Data		77
	3.1	Site description and data handling preliminaries: An AmeriFlux site	77
	3.2	Observed values of the cospectral parameters f_x, Δf_x, α, and $\Delta \alpha$	79
	3.3	Cospectral (dis)similarity between sites: A CarboEurope flux site	84
	3.4	Departures from smooth cospectral shapes	87
	3.5	Results for an open-path system	88
	3.6	Extension to a closed-path system and $\Delta \tau_e$	89
	3.7	Discussion and caveats concerning low frequencies	92
	3.8	Summary	94
4	References		96

5 Low Frequency Atmospheric Transport and Surface Flux Measurements   101

Yadvinder Malhi, Keith McNaughton, Celso Von Randow

1	Introduction		101
2	Turbulence Structure, Eddy Sizes and Sampling Times		102
	2.1	Mixed-layer (outer-layer) timescales	105
	2.2	Surface-layer (inner-layer) timescales	108
3	Empirical Evidence of Low Frequency Flux Transport		111
	3.1	Wavelet spectral analyses of turbulent fluxes in Scotland and Amazonia	111
	3.2	Low frequency transport and energy balance	114
4	Effect of Standard Flux Calculations on Low Frequency Flux Terms		115
5	Complications		116

	6	Conclusions	117
	7	References	117

6 Measurements of Trace Gas Fluxes in the Atmosphere Using Eddy Covariance: WPL Corrections Revisited   119
Ray Leuning

1	Introduction		119
2	Conservation Equations for Moist Air and Trace Constituents		120
3	Non-steady, Three Dimensional Flow		122
4	Steady, One-dimensional Horizontally Homogeneous Flows		125
	4.1	Fluxes	125
	4.2	The vertical velocity of air	126
5	Practical Considerations		127
	5.1	Fluxes in terms of mixing ratios and concentrations	127
	5.2	Closed-path analyzers	127
	5.3	Open-path gas analyzers	129
	5.4	Advection	130
6	Conclusions		130
7	References		131

7 Concerning the Measurement of Atmospheric Trace Gas Fluxes with Open- and Closed-path Eddy Covariance System: The WPL Terms and Spectral Attenuation   133
William Massman

1	Introduction		134
2	The WPL80 Terms and Spectral Attenuation		136
3	Open-path Systems		138
4	Closed-path Systems		140
	4.1	General considerations	140
	4.2	Detection chamber transfer functions	142
	4.3	Pressure fluctuations within the detection chamber	147
	4.4	Synthesis: Possible consequences for flux estimates	150
	4.5	Low frequency temperature fluctuations	155
5	Summary and Conclusions		156
6	Acknowledgment		157
7	References		158

8 Stationarity, Homogeneity, and Ergodicity in Canopy Turbulence   161
Gabriel Katul, Daniela Cava, Davide Poggi, John Albertson, Larry Mahrt

1	Introduction	161
2	Stationarity, Homogeneity, and the Ergodic Hypothesis	164
3	Stationarity, Homogeneity, and Ergodicity in Atmospheric Surface Layer Flows	167
4	Homogeneity and Ergodicity in the Neutral and Unstable CSL	171
5	Stationarity and Ergodicity in the Stable CSL	174
6	Conclusions	176
7	Acknowledgment	178

	8	References	179

9 Post-field Data Quality Control — 181
Thomas Foken, Mathias Göckede, Matthias Mauder, Larry Mahrt, Brian Amiro, William Munger

	1	Introduction		181
	2	Quality Assurance and Quality Control		182
	3	Quality Control of Eddy Covariance Measurements		184
		3.1	Basic tests of the raw data	184
		3.2	Statistical tests	187
		3.3	Tests on fulfillment of theoretical requirements	189
		3.4	Overall quality flag system	193
		3.5	Site dependent quality control	195
	4	Further Problems of Quality Control		198
	5	Conclusion		202
	6	Acknowledgment		203
	7	References		203

10 Advection and Modeling — 209
John Finnigan

	1	Introduction		209
	2	General Remarks on Modeling and Advection		211
		2.1	Modeling	211
		2.2	Advection	212
	3	The Turbulent Wind Field in a Tall Canopy on a Low Hill		218
	4	Scalar Flow and Transport in a Tall Canopy on a Low Hill		226
		4.1	Analytical model	226
		4.2	Numerical model	233
	5	Discussion and Conclusion		238
	6	Appendix A: Model for Stomatal Conductance		240
	7	Appendix B: Model for Photosynthetically Active Radiation		241
	8	Acknowledgment		241
	9	References		241

Index — 245

Contributing Authors

John Albertson
Department of Civil and Environmental Engineering, Duke University, Durham NC 27708-0287, U. S. A.

Brian Amiro
Canadian Forest Service, Northern Forestry Centre, 5320-122 Street, Edmonton, Alberta T6H 3S5, Canada

Daniela Cava
CNR, Institute of Atmosphere Sciences and Climate, Strada Prov. Lecce-Monteroni km 1,200, Polo Scientifico dell'Universitá, 73100 Lecce, Italy

Robert Clement
School of GeoSciences, University of Edinburgh, Edinburgh EH9 3JU, U. K.

John Finnigan
CSIRO Atmospheric Research, P O Box 1666, Canberra ACT 2601, Australia

Thomas Foken
University of Bayreuth, D-95440 Bayreuth, Germany

Mathias Göckede
University of Bayreuth, D-95440 Bayreuth, Germany

Gabriel Katul
Nicholas School of the Environment and Earth Sciences, Duke University, Durham, NC 27708-0328, U. S. A.

Beverly Law
College of Forestry, Oregon State University, Corvallis, OR 97331, U. S. A.

Xuhui Lee
School of Forestry and Environmental Studies, Yale University, New Haven, CT 06511, U. S. A.

Ray Leuning
CSIRO Atmospheric Research, PO Box 1666, Canberra, ACT 2601, Australia

Tilden Meyers
NOAA/ATDD, PO Box 2456, 456 S Illinois Ave, Oak Ridge, TN 37831, U. S. A.

Larry Mahrt
Oregon State University, College of Oceanic and Atmospheric Sciences, Corvallis OR 97331-5503, U. S. A.

Yadvinder Malhi
School of Geography and the Environment, University of Oxford, Mansfield Road, Oxford OX1 3TB, U. K.

William Massman
USDA Forest Service, 240 West Prospect Rd, Fort Collins, CO 80526, U. S. A.

Matthias Mauder
University of Bayreuth, D-95440 Bayreuth, Germany

Keith McNaughton
School of GeoSciences, University of Edinburgh, Edinburgh EH9 3JU, U. K.

John Moncrieff
School of GeoSciences, University of Edinburgh, Edinburgh EH9 3JU, U. K.

Contributing Authors

William Munger
Department of Earth and Planetary Sciences, Harvard University, Cambridge MA 02138, U. S. A.

Kyaw Tha Paw U
Atmospheric Science, University of California, Davis CA 95616-8627, U. S. A.

Davide Poggi
Nicholas School of the Environment and Earth Sciences, Duke University, Durham NC 27708-0328, U. S. A.

Celso von Randow
Alterra Green World Research, Wageningen, The Netherlands

Shashi Verma
School of Natural Resources Science, University of Nebraska, Lincoln NE 68583, U. S. A.

Preface

Micrometeorology is a branch of meteorology that is concerned with atmospheric phenomena and processes near the ground at scales of tens of meters to several kilometers. Progress in micrometeorology is made through experimental investigation of these phenomena and quantitative study attempting to bring order to experimental data. Studies of surface-air flux play a crucial role in this endeavor.

The current paradigm of micrometeorology builds on two premises: (i) that scale separation exists so that the microscale phenomena can be treated more or less in isolation of phenomena occurring at larger scales, and (ii) that these phenomena are influenced by the surface to such an extent that "external factors" can be ignored. Quantitative studies have been based on the assumption of horizontal homogeneity, which inevitably biases the investigation toward over-idealization of the real world by restricting it to perfectly flat topography and daytime, fair weather conditions. This bias was noted by John Philip 40 years ago: "Experimenters attempt to avoid [advection] by working on sites downwind of extensive 'homogeneous' areas. Sometimes advection is invoked to explain otherwise inexplicable observations..." (*J. Meteorol.* **16**, 535).

The international networks of flux sites (FLUXNET) consortium deploy the micrometeorological methods as core methodology to achieve the goal of increasing our understanding of energy and mass exchange between the biosphere and the atmosphere. The FLUXNET scientists have produced a growing body of experimental evidence that demonstrates the deficiencies of the "flat-earth" paradigm. Their work also highlights that, without a uniform theoretical framework, execution of a field program and the subsequent data interpretation could be subject to considerable confusion. It is against this context that an international workshop was hosted by AmeriFlux in 2002 to discuss standardization of flux diagnostics and analysis. This book volume is a collection of writings by the workshop invitees, on topics we believe most relevant to the surface flux observation and diagnostics. It is our hope that the book will bring some coherence to estimates of mass and energy exchange and will simulate efforts to study phenomena that may fall outside the scope of the normal science of micrometeorology.

We thank Roger Dahlamn, U. S. Department of Energy, for his support for the workshop and his continued encouragement that have made this book volume possible. We would like to acknowledge the invited lecturers and discussants, whose names appear in Chapter 1, for their participation in the workshop and the subsequent writing assignment. We are also grateful to Peter Anthoni, Dave Billesbach, Constance Brown-Mitic, George Burba, Matthias Falk, Chris Fiebrich, Marc Fischer, Larry Hipps, Jinkyu Hong, John Hunt, Joon Kim, Meredith Kurpius, Chun-Ta Lai, Monique Leclerc, Hank Loescher, Kai Morgenstern, John Nagy, Elizabeth Pattey, Ruth Reck, Dan Ricciuto, Russell Scott, Julie Styles, Andy Suyker, Susan Ustin, Shashi Verma, Dean Vickers and Marvin Wesely, who contributed to the workshop discussion and debate.

XUHUI LEE

WILLIAM MASSMAN

BEVERLY LAW

Chapter 1

INTRODUCTION

Beverly Law, Shashi Verma
bev.law@oregonstate.edu

Abstract This book summarizes and extends the discussion at an international workshop on eddy covariance flux analysis and diagnostics. Its goal is to provide micrometeorologists, ecosystem scientists, boundary-layer meteorologists, and students involved in micrometeorology with the state of science on measurement and analysis of exchange of mass and energy between the terrestrial biosphere and the atmosphere. It provides useful advice for bringing coherence to estimates of mass and energy exchange and for cross-site comparisons and synthesis activities.

Within the global network of micrometeorological tower sites (FLUXNET) there are several regional networks that are making micrometeorological measurements to quantify and understand the spatial and temporal variations in carbon storage in plants and soils, and the exchanges of carbon dioxide, water vapor, and energy in major vegetation types (e. g. grasslands, agricultural crops, tropical forests, temperate coniferous and deciduous forests) across a range of disturbance histories and climatic conditions. They include AmeriFlux, CarboEurope, FLUXNET-Canada, OzFlux, AsiaFlux. In these networks the exchanges of carbon dioxide, water vapor and energy are measured employing the micrometeorological eddy covariance technique (e. g., Baldocchi et al. 1988). The eddy covariance technique provides a relatively direct means of measuring fluxes, without the need for assumptions regarding eddy diffusivities. Early eddy covariance CO_2 flux studies were limited to field campaigns in agricultural crops (e. g., Anderson et al. 1984, Desjardins 1985), grasslands (e. g., Verma et al. 1989), and forests (e. g., Verma et al. 1986, Baldocchi and Meyers 1991, Valentini et al. 1991, Hollinger et al. 1994) over short periods. With further technological advances, year-round flux measurements became feasible in early 1990s (e. g., Wofsy et al. 1993, Black et al. 1996, Greco and Baldocchi 1996, Anthoni et al.

1999, Aubinet et al. 2001). It is, however, worthwhile to keep in mind that a long term operation of eddy covariance sensors presents a number of challenges, including appropriate maintenance and calibration of sensors and data acquisition equipment.

The existence of these networks provides an unprecedented opportunity to conduct cross-site comparisons and synthesis studies in a range of terrestrial ecosystems. To examine site-to-site differences in fluxes of mass and energy, we need to understand and reduce uncertainties in flux estimates, and develop appropriate data QA/QC and archiving procedures. It is important to develop protocols to minimize the impact of differences in sensors and data processing procedures. Accordingly, a workshop, co-chaired by W. J. Massman and X. Lee, was sponsored by NIGEC (National Institute for Global Environmental Change) to address these issues on 30-31 May 2000 in Boulder, Colorado, USA (Massman and Lee 2002). A follow-up workshop was sponsored by the U. S. Department of Energy on 27-30 August 2002 at Oregon State University (Corvallis, Oregon, USA), and was attended by representatives from the different flux networks.

The Corvallis workshop covered the following topics, which are considered critical to the long-term objectives of the flux networks. These are:

- Averaging and Detrending (J. Moncrieff, lecturer; T. Meyers, discussant)

- Coordinate Rotation (X. Lee, lecturer; K. T. Paw U, discussant)

- Low Frequency Corrections (Y. Mahli, lecturer; D. Baldocchi, discussant)

- High Frequency Corrections (W. J. Massman, lecturer; R. Clement, discussant)

- Flux Corrections for Cross Contamination (R. Leuning, lecturer; S. Miller, discussant)

- Time Series Analysis (G. Katul, lecturer; L. Mahrt, discussant)

- Post-field Data Quality Controls (T. Foken, lecturer; B. Amiro and W. Munger, discussants)

- Advection and Modeling (J. Finnigan, lecturer; B. Heinesch and H. P. Schmid, discussants)

The chapters in this book summarize the key topics and recommendations for flux data analysis. Chapter 2 reviews main averaging and detrending methods to produce fluctuations and means. Flux loss associated with these methods is illustrated with data from a number of FLUXNET sites. It suggests that block averaging is usually the best option. To determine the optimal averaging period at a site, they suggest the ogive method.

Chapter 3 examines theoretical and operational aspects of coordinate systems, including an assessment of flux bias errors due to sensor tilt in horizontally homogeneous flow. Workshop participants agreed that application of the planar fit coordinate rotation procedure is the preferred method.

Chapter 4 addresses the uncertainties in eddy covariance flux measurements that have been corrected for spectral attenuation with the transfer function approach. The sources of error in the estimates of flux attenuation are discussed and a method is proposed for estimating the uncertainty in measured covariance.

Chapter 5 examines the contributions of low frequency fluctuations to fluxes. A workshop recommendation was that the flux averaging period should be no shorter than 30 and no greater than 60 minutes. However, longer averaging periods will be required to fully investigate low frequency contributions to the fluxes.

Chapter 6 re-examines the Webb, Pearman, and Leuning (1980: WPL) corrections associated with the calculation of trace gas fluxes using the eddy covariance technique. It was concluded that theory developed by WPL for one-dimensional flows is still applicable for the vertical component of eddy fluxes and the equations are simplified when gas concentrations are expressed as mixing ratios per unit of dry air. Chapter 7 offers a complementary perspective on the WPL theory by examining how open- and closed-path infrared gas analyzers used for eddy flux measurements influence the application of this theory.

Chapter 8 provides a review of the stationarity and ergodicity concepts (two required conditions) used to link field measurements and the Navier-Stokes equations of motion or field measurements to boundary conditions at the land-atmosphere interface. The concepts are reviewed for the atmospheric surface layer and canopy sublayer turbulence, and the authors show how the stable canopy sublayer tends to violate both conditions. Practical fixes such as thresholds based on the friction velocity (u_*) to correct nighttime CO_2 fluxes are shown to be a reasonable starting point for dealing with these issues.

Chapter 9 summarizes quality assurance and quality control procedures for eddy covariance measurements. The authors address electronic (instrument related), meteorological and statistical issues, including how closely conditions fulfill the theoretical assumptions underlying the eddy covariance method. Also addressed are procedures for data quality analysis using footprint models. They describe a set of possible tests and protocols for flagging data and provide practical advice for use in continuously running eddy covariance systems.

Recent modeling studies and field experiments show that horizontal and vertical advection terms tend to be of opposite sign and comparable magnitude. However, in complex terrain, their sum is not necessarily zero and it can make a significant contribution to the calculation of surface exchange. Chapter 10 discusses advection conditions and the state of our knowledge, although it is premature to make general recommendations for operational corrections for advection. A concerted measuring and modeling effort with site intercomparisons will be needed.

Also covered at the workshop, but not explicitly included here, were sessions on software development and consensus building and discussions of emerging scientific issues. These latter discussions tended to be associated with the influence of advection or complex terrain on measured fluxes. The topics included drainage flows, wind shear inside canopies, tower flux measurement height, and inspection of turbulent time series to diagnose consequences of low wind conditions on flux data. The exchange and testing of software was also encouraged within the community to ensure consistency and to minimize redundant efforts.

This Handbook of Micrometeorology is intended to provide micrometeorologists, ecosystem scientists, boundary-layer meteorologists, and students involved in micrometeorology with the state of science on measurement and analysis of exchange of mass and energy between the terrestrial biosphere and the atmosphere. It is the culmination of many detailed discussions of theory, analysis, and practical applications, with the expectation that it will provide useful advice for bringing coherence to estimates of mass and energy exchange, which is essential for cross-site comparisons and synthesis activities.

References

Anderson, D. E., Verma, S. B., Rosenberg, N. J.: 1984, 'Eddy correlation measurements of CO_2, latent heat and sensible heat fluxes over a crop surface', *Boundary-Layer Meteorol.* **29**, 167-183.

Anthoni, P. M., Law, B. E., Unsworth, M. H.: 1999, 'Carbon and water vapor exchange of an open-canopied ponderosa pine ecosystem', *Agric. For. Meteorol.* **95**, 151-168.

Aubinet, M., Chermanne, B., Vandenhaute, M., Longdoz, B., Yernaux, M., Laitat, E.: 2001, 'Long term carbon dioxide exchange above a mixed forest in the Belgian Ardennes', *Agric. For. Meteorol.* **108**, 293-315.

Baldocchi, D., Meyers, T. D.: 1991, 'Trace gas exchange above the floor of a deciduous forest. 1. Evaporation and CO_2 efflux', *J. Geophys. Res.* **96**, 7271-7285.

Baldocchi, D., Hicks, B. B., Meyers, T. D.: 1988, 'Measuring biosphere-atmosphere exchanges of biologically related gases with micrometeorological methods', *Ecology* **69**, 1331-1340.

Black, T. A., Den Hartog, G., Neumann, H. H., Blanken, P. D., Yang, P. C., Russell, C., Nesic, Z., Lee, X., Chen, S. G., Staebler, R., Novak, M. D.: 1996, 'Annual cycles of water vapor and carbon dioxide fluxes in and above a boreal aspen forest', *Global Change Biol.* **2**, 219-229.

Desjardins, R. L.: 1985, 'Carbon dioxide budget of maize', *Agric. For. Meteorol.* **36**, 29-41.

Greco, S., Baldocchi, D.: 1996, 'Seasonal variations of CO_2 and water vapor exchange rates over a temperate deciduous forest', *Global Change Biol.* **2**, 183-197.

Hollinger, D. Y., Kelliher, F. M., Byers, J. N., Hunt, J. E., McSeveny, T. M., Weir, P. L.: 1994, 'Carbon dioxide exchange between an undisturbed old-growth temperate forest and the atmosphere', *Ecology* **75**, 134-150.

Massman, W. J., Lee, X.: 2002, 'Eddy covariance flux corrections and uncertainties in long-term studies of carbon and energy exchanges', *Agric. For. Meteorol.* **113**, 121-144.

Valentini, R., Scarascia Mugnozza, G. E., De Angelis, P., Bimbi, R.: 1991, 'An experimental test of the eddy correlation technique over a Mediterranean macchia canopy', *Plant Cell Environ.* **14**, 987-994.

Verma, S. B., Kim, J., Clement, R.: 1989, 'Carbon dioxide, water vapor and sensible heat fluxes over a tallgrass prairie', *Boundary-Layer Meteorol.* **46**, 53-67.

Verma, S. B., Baldocchi, D. D., Anderson, D. E., Matt, D. R., Clement, R. J.: 1986, 'Eddy fluxes of CO_2, water vapor and sensible heat over a deciduous forest', *Boundary-Layer Meteorol.* **36**, 71-91.

Webb, E. K., Pearman, G. I., Leuning, R.: 1980, 'Correction of flux measurements for density effects due to heat and water vapour transfer', *Q. J. Royal Meteorol. Soc.* **106**, 85-100.

Wofsy, S. C., Goulden, M. L., Munger, J. W., Fan, S.-M., Bakwin, P. S., Daube, B. C., Bassow, S. L., Bazzaz, F. A.: 1993, 'Net exchange of CO_2 in a mid-latitude forest', *Science* **260**, 1314-1317.

Chapter 2

AVERAGING, DETRENDING, AND FILTERING OF EDDY COVARIANCE TIME SERIES

John Moncrieff, Robert Clement, John Finnigan, Tilden Meyers
moncrieff@ed.ac.uk

Abstract   Data from sensors in an eddy covariance system are routinely processed to remove trends and to produce fluctuations and means. Historically this has been seen to be a relatively straightforward task and the methods are well known. Such re-processing can result in the loss of real signal since the detrending and averaging methods act as high-pass filters. We review the main methods used to separate the active, turbulent transport that we treat as eddy flux from the slower, deterministic atmospheric motions and instrument drift. We discuss the advantages and disadvantages of various algorithms used in averaging, detrending and filtering and conclude that the best method is likely to be dependent on site conditions and data processing system in use. We recommend the use of the ogive to determine the optimal averaging period at any site. We illustrate outstanding issues with data from a number of FLUXNET sites.

1   Introduction

Eddy covariance is the predominant method in FLUXNET and has a-chieved its popularity because of the relative robustness of both its theoretical underpinnings and modern environmental sensors. Having said that, it is unlikely that the conditions of stationarity in time and homogeneity in space under which the original theories were established are ever strictly met in practice. This is coming into sharp focus because many FLUXNET sites are in areas of complex terrain, where flow is inhomogeneous and because, in attempting to measure 24 hours a day 365 days a year, we encounter non-stationarity on a regular basis.

Micrometeorological data-processing techniques in the past have been predicated on ideal flow conditions and so must be carefully re-evaluated for the more challenging conditions of FLUXNET. Some of these techniques such as coordinate rotation, data quality checking and instrument corrections are covered in later chapters in this volume. Here, we will concentrate on the averaging, detrending, and filtering operations used to separate the turbulent signals that are to be included in the eddy flux from trends or low frequency components imposed either by instrumental drift or as a result of changes in meteorological conditions.

Hitherto, a starting point for discussing filtering and averaging has been to assume that dealing with trends in measured data was similar to dealing with finite length samples taken from ideal time series of infinite length and zero ensemble mean. Variances, covariances and higher moments obtained from these finite-length samples could then be compared with the expectations of the same quantities over the infinite ensemble. Such comparisons allowed the effect on chosen statistics of operations like mean removal, detrending over an averaging period T or high-pass filtering to be evaluated and presented as corrections, quantifying the effect of the chosen operation on the ideal statistics (e. g. Lenschow et al. 1994, Rannik and Vesala 1999).

While these comparisons continue to provide useful quantitative bounds on the effect of such operations, in processing flux tower data we cannot assume that what we measure is a departure from some ideal stationary time series. Instead we must accept that the eddy flux is simply that part of the mass, momentum or energy transport that is carried by turbulent motions in the planetary boundary layer (PBL). These motions are part of a continuous spectrum of atmospheric fluctuations with time scales from seconds to seasons and length scales from meters to kilometers and beyond. On any practical time scale these series are intrinsically non-stationary. Our task here is to discuss the techniques that allow a rational separation of the transport process into the strong and active part that we identify as eddy fluxes and deal with by a variety of statistical techniques, and slower deterministic processes. A good starting point is to review briefly the basic assumptions of the eddy flux method, which might more accurately be termed the *aerodynamic method* of measuring surface exchange.

The basis of the aerodynamic method is to erect a notional control volume over a representative patch of surface, to measure the exchange across all the aerial faces of this volume, as well as recording any accumulation within it, and then to infer the surface exchange by difference. We rely on turbulent mixing to act as a *physical averaging operator* so that measurements at some height h capture exchange from a representative surface patch. If we assume that, when averaged over a sufficiently long time, the flow field is effectively one-dimensional, we can write,

$$\frac{\overline{\partial c}}{\partial t} + \frac{\partial \overline{wc}}{\partial z} = \overline{S}\delta(z) \tag{2.1}$$

where $c(t)$ is a generic scalar, $w(t)$ is the vertical component of the velocity vector, the overbar denotes a filtering, detrending or averaging operation, $\overline{S}$ is the surface exchange, z the vertical or surface normal coordinate and $\delta(z)$ is the Dirac delta function. A more rigorous account of the steps and assumptions leading from the full, 3-dimensional, non-stationary situation to Equation 2.1 can be found in Chapter 10 and in Finnigan et al. (2003). When the flow field is stationary, i. e. when there is no accumulation of c in the notional control volume, then the first term on the left hand side of Equation 2.1 is zero and the equation reduces to $\partial \overline{wc}/\partial z = \overline{S}\delta(0)$.

Next, integrating Equation 2.1 from the ground at $z = 0$ to the sensor at height h, the top of the control volume, we have

$$\overline{wc}(h) = \overline{S} \tag{2.2}$$

The left hand side of Equation 2.2 is the total covariance of $w(t)$ and $c(t)$ under the chosen averaging operator. One final step is required to replace $\overline{wc}(h)$ by the measured eddy flux. This is to separate the slowly varying 'background' variations in $w(t)$ and $c(t)$, which we write as $\overline{w}(t)$ and $\overline{c}(t)$, from the rapid turbulent variations about $\overline{w}(t)$ and $\overline{c}(t)$. Without specifying the averaging operator[1] we write,

$$w(t) = \overline{w}(t) + w'(t) \tag{2.3}$$

and

$$c(t) = \overline{c}(t) + c'(t) \tag{2.4}$$

whence,

$$\overline{wc} = \overline{\overline{w}\,\overline{c}} + \overline{\overline{w}c'} + \overline{w'\overline{c}} + \overline{w'c'} \tag{2.5}$$

If the averaging operator obeys the desirable Reynolds averaging properties, then Equation 2.5 reduces to,

$$\overline{wc} = \overline{\overline{w}\,\overline{c}} + \overline{w'c'} \tag{2.6}$$

---

[1] Henceforth, for brevity we take the term 'averaging' to include filtering and detrending unless otherwise stated.

Finally, in horizontally homogeneous flows with z normal to the surface, $\overline{w} \to 0$ (see Chapter 3) and the integrated mass balance, Equation 2.2 becomes a statement of the equality of the eddy flux at height h and the surface exchange,

$$\overline{w'c'}(h) = \overline{S} \tag{2.7}$$

We have gone through the foregoing steps in detail to emphasize that the eddy flux is equal to the surface exchange only under a set of quite restrictive conditions. Horizontal homogeneity is necessary if we are to ignore horizontal flux divergences, stationarity is required to ignore the storage term, and both coordinate rotation and an averaging operator that obeys Reynolds averaging rules are required to replace the total covariance $\overline{wc}$ by the eddy covariance, $\overline{w'c'}$.

In this Chapter we will address four questions:

- How do we decide what motions to count as eddy flux and what as slow variations to be treated deterministically or as instrumental drift to be discarded?

- What are the best tools to separate slowly changing (mean) and rapidly varying (turbulent) parts of our time series?

- Can we distinguish between instrument drift and low frequency meteorological signal in the data?

- If we cannot distinguish, how large an error do we make if we discard some low frequency meteorological signal along with instrument drift?

The first question that a practical experimenter should ask, of course, is how grave a sin is committed if the suite of questions listed above is ignored. How large is the error if we treat measurements as if they were made in ideal conditions, even when we know they were not? Table 2.1 shows an analysis by Moors et al. (poster presented at CarboEurope Conference, Budapest, March 2002) of the sensitivity of annual carbon totals at their flux sites in the Amazon to individual error terms in their flux system. Their conclusion was that the most important term was 'rotation and averaging', giving rise to 10-25% uncertainty in their estimates. The other terms associated with the hardware were of lesser importance. The implication is clear; we need to re-evaluate the procedures that we regarded as simple, almost unexceptional in the past.

Table 2.1. Uncertainty estimates on CO_2 flux, F_c, for a typical eddy covariance system from Moors et al. (poster presented at CarboEurope Conference, Budapest, March 2002). Here $\sigma(d)$ is the uncertainty in the zero-plane displacement, $\sigma(f_t)$ is the uncertainty in tube flow rate (f_t), n_{cf} is the number of cycles in tube flow rate, and n_d is number of days used to fit a fill relationship. For Manaus and Jaru towers, $\sigma(d) = 10$ m, $\sigma(f_t)/f_t = 0.5$, $n_{cf} = 4$ y^{-1}.

	Systematic error	Random error on half-hour F_c	Total one-sided error on ann. sum
Spikes/noise	2%	11%	2%
Tube delay	–	3.5%	< 0.1%
Rotation/averaging	10 – 25%	–	10– 25 %
Freq. loss correction (zero plane)	0.27%×$\sigma(d)$	–	2.7%
Freq. loss correction (flow rate)	1%×$\sigma(f_t)/(f_t n_{cf}^{0.5})$	–	<0.5%
Missing data filling	–	0.08-1.0 or 30-50/$n_d^{1/2}$ kg C ha^{-1} d^{-1}	0.25-1 t C ha^{-1} y^{-1} or 3%–20%

2 Averaging, Detrending, and Filtering Operations

All the operations we perform are in either the time or frequency domains. Although spatial averaging implicitly underlies the eddy flux method and ensemble averaging underpins much basic theory, we cannot actually apply either operator to measurements made at a single tower. It is useful at the outset, therefore, to clarify the effect in the two domains of the three main types of operation available to us, time averaging, detrending, and filtering.

2.1 Time averaging

As a starting point we assume that we have effectively infinite (very long anyway) time series $w(t)$ and $c(t)$. We want to divide these series into consecutive segments of length T and average them over this period. The time average operator is defined as,

$$\overline{w(t)} = \overline{w} = \frac{1}{T}\int_0^T w(t)\,dt \qquad (2.8)$$

and we remove this mean in the period T to define the turbulent fluc-

tuation,
$$w'(t) = w(t) - \overline{w} \qquad (2.9)$$

The covariance or eddy flux in the period T is then defined as,
$$\overline{w'c'} = \overline{(w(t) - \overline{w})(c(t) - \overline{c})} \qquad (2.10)$$
and because $\overline{w}$ and $\overline{c}$ are constants we can write,
$$\overline{wc} = \overline{w}\,\overline{c} + \overline{w'c'} \qquad (2.11)$$
so time averaging with mean removal (MR) obeys the Reynolds averaging conditions. The effect of time averaging and mean removal on the cospectra is somewhat involved but it can be approximated quite well by the running mean filter operation discussed in Section 2.3 below, so long as the averaging time T is much longer than the period of any fluctuations in the original time series, $w(t)$ and $c(t)$. (Strictly we require $\tau/T << 1$, where τ is the integral time scale of the turbulent time series; Kristensen 1998).

In Equations 2.8 to 2.11 we have treated the signals as smooth functions but in reality of course we deal with digitally sampled signals so that the time averaging and mean removal operations are performed in discrete form over the n_s samples of these signals in the averaging period T,

$$\overline{w} = \frac{1}{n_s} \sum_{k=1}^{n_s} w_k(t) \qquad (2.12)$$
$$w'_k(t) = w_k(t) - \overline{w}$$

and compute the eddy covariance of $w(t)$ and $c(t)$ as

$$\overline{w'c'} = \frac{1}{n_s} \sum_{k=1}^{n_s} w'_k c'_k \qquad (2.13)$$

The effect of mean removal on a typical turbulent signal is illustrated in Figure 2.1a.

2.2 Linear detrending

In linear detrending, instead of subtracting the mean from the signal in a period T we find the line of best fit over the period, i. e. the linear trend, and subtract that. This is illustrated in Figure 2.1b. Writing the line of best fit to $w(t)$ in period T as,

$$W(t) = W_I + W_S t$$

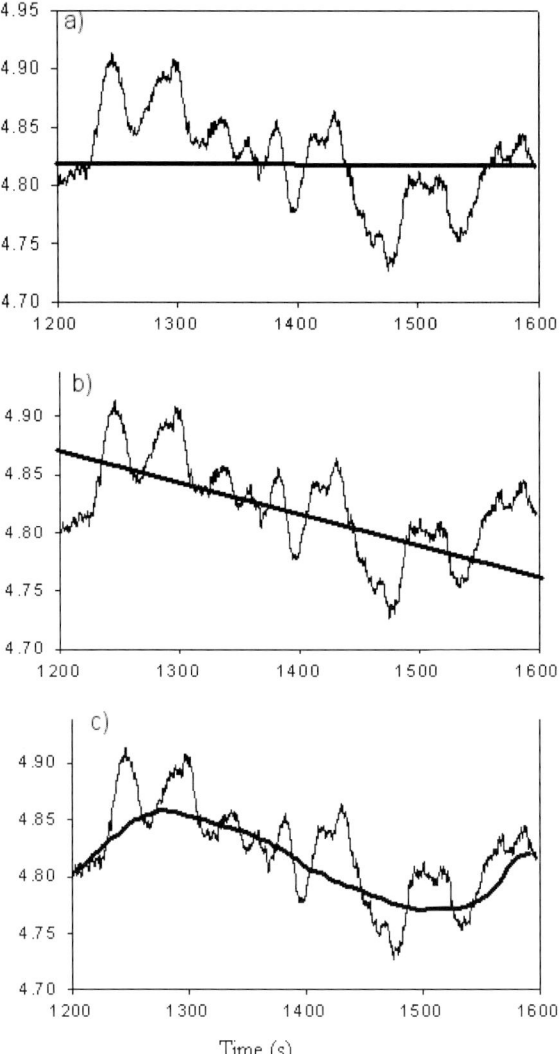

Figure 2.1. Illustration of detrending by three different algorithms on the same data series. a) block average, b) linear detrend, c) recursive. The y-axis is an arbitrary scale.

where W_I is its intercept and W_S its slope, we now define,

$$w'(t) = w(t) - W(t) \qquad (2.14)$$
$$c'(t) = c(t) - C(t)$$

where $C(t)$ represents the line of best fit to $c(t)$. The covariance or eddy flux in the period T is now defined as,

$$\overline{w'c'}(t) = \overline{\left(w(t) - \overline{W(t)}\right)\left(c(t) - \overline{C(t)}\right)} \quad (2.15)$$

and because $W(t)$ and $C(t)$ are not constant we must write,

$$\overline{wc}(t) = \overline{W(t)C(t)} + \overline{W(t)c'(t)} + \overline{w'(t)C(t)} + \overline{w'c'} \quad (2.16)$$

In other words, linear detrending does not obey Reynolds averaging rules. The reason is that the subtraction of the linear trend in the time domain is equivalent in the frequency domain to subtracting the Fourier transform of $W(t)$ and $C(t)$ from the spectra of $w(t)$ and $c(t)$. $W(t)$ and $C(t)$ are ramp or saw-tooth functions. Their transforms decay with frequency ω like $1/|\omega|^2$ and so have contributions at all frequencies. Hence, linear detrending while primarily affecting the low frequency part of the signal, affects all frequencies.

To apply the linear detrend to sampled data $\overline{c_k}$ we use least squares regression (Gash and Culf 1996),

$$\overline{c_k} = \overline{c} + b\left(t_k - \frac{1}{n_s}\sum_{k=1}^{n_s} t_k\right) \quad (2.17)$$

where t_k is the time at step k and b is the slope of the line of best fit to the sample, computed by:

$$b = \frac{\sum_{k=1}^{n_s} c_k t_k - \frac{1}{n_s}\sum_{k=1}^{n_s} c_k \sum_{k=1}^{n_s} t_k}{\sum_{k=1}^{n_s} t_k t_k - \frac{1}{n_s}\sum_{k=1}^{n_s} t_k \sum_{k=1}^{n_s} t_k} \quad (2.18)$$

2.3 Filtering

Filtering is defined as convolution of the signal $w(t)$ or $c(t)$ in the time domain with a window function $G(t)$. In the frequency domain this is equivalent to multiplying the spectrum of the unfiltered signal by the Fourier transform of the window. Hence the transfer function of a filter is just the Fourier transform of its window shape in the time domain. Neither averaging with mean removal (MR) nor linear detrending (LDT) are true filtering operations as they involve subtraction in the time and frequency domains rather than convolution. Denoting the low

Averaging, Detrending, and Filtering

Table 2.2. The three main high-pass filters and their transfer functions (1, block average or mean-only removal; 2, linear detrend; 3, running mean). Here A, B, C, D are coefficients of the linear regression, f is frequency, Δt is sampling interval, τ_f is the RC filter time constant, and T is averaging length.

Filter	Filter algorithm	Transfer function
1	$\overline{w's'} = \dfrac{1}{n}\sum ws - \dfrac{1}{n^2}\sum w \sum s$	$\dfrac{\sin^2(\pi fT)}{(\pi fT)^2}$
2	$\overline{w's'} = \dfrac{1}{n}\sum_i (w - [A + Bi])(s - [C + Di])$	$1 - \left[\dfrac{\sin^2(\pi fT)}{(\pi fT)^2} - 3\dfrac{\left(\dfrac{\sin(\pi fT)}{\pi fT} - \cos(\pi fT)\right)^2}{(\pi fT)^2}\right]$
3	$s'_t = s_t - a\overline{s}_{t-\Delta t} + (1-a)s_t$	$\dfrac{(2\pi f\tau_f)^2}{1 + (2\pi f\tau_f)^2}$
	where $a = e^{-\Delta t/\tau_f}$	

pass filtered part of the signal by $\tilde{w}(t)$,

$$\tilde{w}(t) = \int_{-T}^{T} G(t' - t)w(t')\,dt' \tag{2.19}$$

where $w(t) = \tilde{w}(t) + w'(t)$ and $c(t) = \tilde{c}(t) + c'(t)$ so that,

$$\overline{w'c'} = \overline{(w(t) - \tilde{w}(t))(c(t) - \tilde{c}(t))} \tag{2.20}$$

Therefore, just as for linear detrending, filtering in general does not obey Reynolds averaging rules and we must write,

$$\overline{wc} = \overline{\tilde{w}\tilde{c}} + \overline{\tilde{w}c'} + \overline{w'\tilde{c}} + \overline{w'c'} \tag{2.21}$$

The shape of the transfer function in the frequency domain depends entirely on the window shape in the time domain (Table 2.2). Unfortunately, simple time windows rarely have transfer functions that provide sharp cut-offs in the frequency domain. A full account of filter shapes is beyond the scope of this Chapter and the reader is referred to standard texts on signal processing such as Bendat and Piersol (1958). However, three filter shapes deserve special mention. The first is the moving average. In this case the window shape is given by,

$$G(t;T) = \begin{cases} 1/2T & \text{for } |t| \leq T \\ 0 & \text{for } |t| > T \end{cases} \tag{2.22}$$

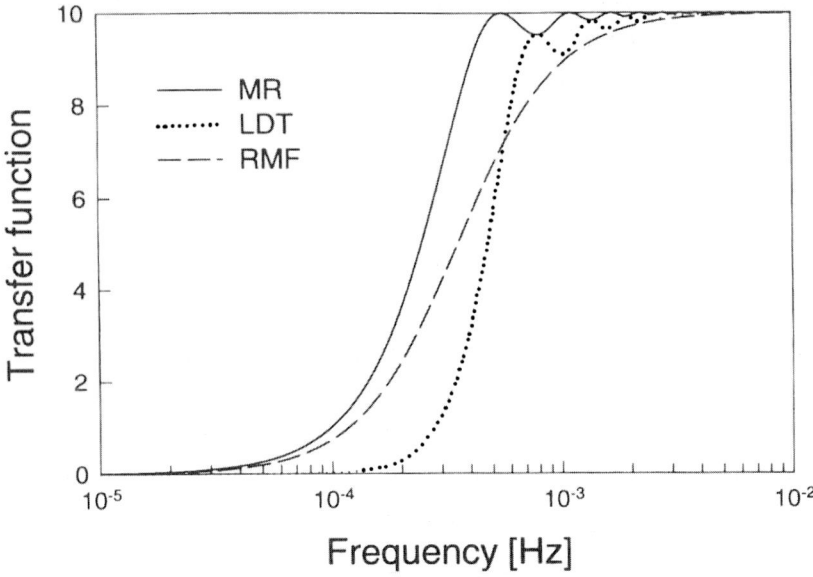

Figure 2.2. Transfer functions for three different algorithms. MR = mean removal; LDT = linear detrend; RMF = running mean filter (recursive). Reproduced with permission of Rannik and Vesala (1999).

Hence every point in the filtered series is just the average of the points in the original series contained in the interval T centered on the current point. The transfer function of the moving average is shown in Figure 2.2 in the form appropriate for use with power or cospectra and, as noted earlier in Section 2.1, this can be taken as an approximation to the effect of mean removal when $\tau/T \ll 1$.

The second filter shape in wide use in eddy flux applications is the recursive digital filter that is an exact analog of a simple, single pole RC filter (Moore 1986, McMillen 1988). This is defined as,

$$\tilde{c}_k = e^{-\Delta t/\tau_f} \tilde{c}_{k-1} + \left(1 - e^{-\Delta t/\tau_f}\right) c_k \qquad (2.23)$$

where Δt is the interval between samples and τ_f is the RC filter time constant. Aubinet et al. (2001) note that when $\Delta t \ll \tau_f$, as is normally the case, Equation 2.23 may also be written after 1st order Taylor expansion as,

$$\tilde{c}_k = \left(1 - \frac{\Delta t}{\tau_f}\right) \tilde{c}_{k-1} + \frac{\Delta t}{\tau_f} c_k \qquad (2.24)$$

Hence the filtered signal at time t (or equivalently index k) is determined only by the present and previously sampled values of the signal, the earlier values having an exponentially decreasing influence on the current value of the filtered signal. This is convenient as the RC filtered signal can be computed continuously as data are recorded rather than having to apply the filter window later to the stored time series. The effect of the RC filter in the time domain is shown in Figure 2.1c.

The third most common filtering operator has until now been applied mainly by accident. The procedure of rotating coordinates so that $\overline{w}$, the mean vertical velocity component in each averaging period T, goes to zero, has the effect of high-pass filtering the time series so that motions of period greater than T make no contribution to the eddy flux while high frequency contributions are distorted (Finnigan et al. 2003). We postpone a discussion of this unconscious filtering until Section 6 below.

2.4 Advantages and disadvantages of the three methods: Their effect on the spectra

The main advantages of time averaging with mean removal is its simplicity and familiarity. Also, the operation obeys Reynolds averaging rules so that the total covariance can easily be reconstructed, if the means have been stored. Similarly, linear detrending is intuitive and, unlike filtering operations, only data in the current time interval are needed to apply the detrend. Unlike the time average, it does not obey Reynolds averaging although in most practical cases the extra so-called Leonard terms in Equation 2.16 $\overline{Wc'} + \overline{w'C}$ are small. Filtering can be less convenient as in most cases it must be applied to a stored time series of raw data although, as shown above, this can be circumvented by using the recursive lagged RC filter. Unlike MR and LDT however, to obtain a filtered record in a period T requires access to a time series longer than T.

We can compare the effects of these three operations in the frequency domain. In Figure 2.2 we have compared the transfer functions corresponding to each of these operations. As noted above in the case of MR and LDT these transfer functions are only approximations to the effect of those operations but they are sufficiently close for this comparison [see Kristensen (1998) for the relationship between the 'pseudo' transfer functions for MR and LDT and the 'true' transfer function of the RC filter.] The time constant of the RC filter is set to 40 s and the time period T of the MR and LDT operations is 1800 s. With these settings, mean removal gives the sharpest cut-off, removing the mean and strongly attenuating lower frequencies. However, it also affects higher frequencies having a decreasing oscillatory form as ω increases. This is because the

Fourier transform of the square window, $G(t,T)$, and hence the transfer function of the moving average is the function $\sin(\omega T/2)/(\omega T/2)$, which has a characteristic oscillatory shape.

Linear detrending removes more low frequencies from the signal as expected but also shows the oscillations at higher frequencies that come from $\sin(\omega T/2)/(\omega T/2)$ as well as removing more of the high frequencies. This is because the Fourier transform of the ramp function has more high frequency content than the square window and so affects the high frequency part of the signal more. Finally the RC filter has the least sharp cut-off but has a much more predictable shape at the high frequency end of the spectrum.

Armed with this survey of the tools at our disposal, we can now ask what time period T or filter time constant τ_f we should choose to separate low frequency variations from turbulence.

3 Choosing an Averaging Period or Filter Time Constant

In the past several decades our community has almost universally adopted time periods of between 10 and 60 minutes as the averaging interval, T over which to calculate means and products. The literature is replete with papers on methods describing techniques and for the special problems of field measurement we have been well-served by accounts such as McMillen (1988), Moore (1986), Lenschow et al. (1994), Rannik and Vesala (1999), and Aubinet et al. (2001) while texts like Kaimal and Finnigan (1994) address the processing of turbulence data more generally. We have had many field campaigns and intercomparison studies that have increased our understanding of land-atmosphere exchange processes. Is there any reason now to question any of these ideas? The answer is certainly yes as recent re-evaluations of the optimum periods for time averaging by Sakai et al. (2001), Finnigan et al. (2003) and Chapter 5 of this book suggest that at many sites, longer averaging periods may be appropriate.

We see this in Figure 2.3 where the carbon dioxide flux, F_c, is calculated over a series of different averaging periods using data from a flux tower site in the Amazon near Manaus (Finnigan et al. 2003). In this case the filtering operation applied was that of coordinate rotation in each averaging period, T, so that the vertical mean velocity over T is set to zero. As discussed in Sections 3 and 6 below, this is roughly equivalent to high-pass filtering the signals so that turbulent motions of period longer than T cannot contribute to the eddy flux. In each panel of Figure 2.3 the ordinate denotes the flux over longer average

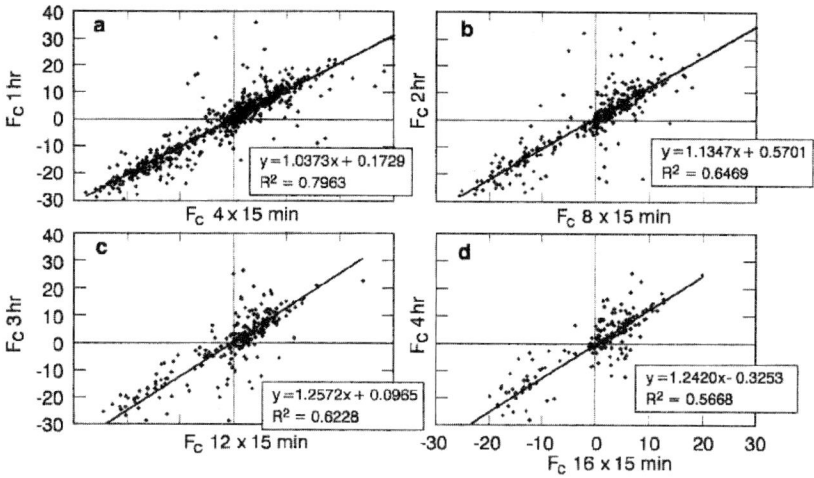

Figure 2.3. Fluxes of carbon dioxide, F_c (μmol m^{-2}s^{-1}), calculated by averaging several consecutive 15-minute periods. Data from Manaus (Finnigan et al. 2003).

periods of 1, 2, 3 and 4 hours respectively. The abscissa is the flux over the same period but constructed from the consecutive 15 min periods that make up each longer period. Hence in panel (a) the ordinate gives F_c, including eddies up to one hour period while the abscissa shows F_c over the same hour but with no eddies longer than 15 min contributing. Clearly eddies with periods between 15 min and 1 hour increase F_c by 3.7% while those between 1 and 2 hours cause a further increase to 13.5%. Almost 10% more flux is carried by eddies between 2 and 3 hour period but little further increase is then observed. Finnigan et al. (2003) applied this comparison to sensible and latent heat and carbon dioxide eddy fluxes at three FLUXNET sites and found that at two of them it was necessary to increase averaging times to as long as 4 hours to capture all the flux while at a third, conventional averaging periods of 30 minutes to an hour were perfectly adequate.

How then do we determine what is an adequate period for our site? Eddy fluxes need to be formed over a sufficiently long time that any motions that contribute to the transport can be sampled adequately. In practice this has meant that eddy fluxes have been calculated over time periods up to an hour in duration, sufficient for several of the largest PBL-scale eddies to be sampled by the measuring system. Periods much longer than this were thought to be inappropriate since signals associ-

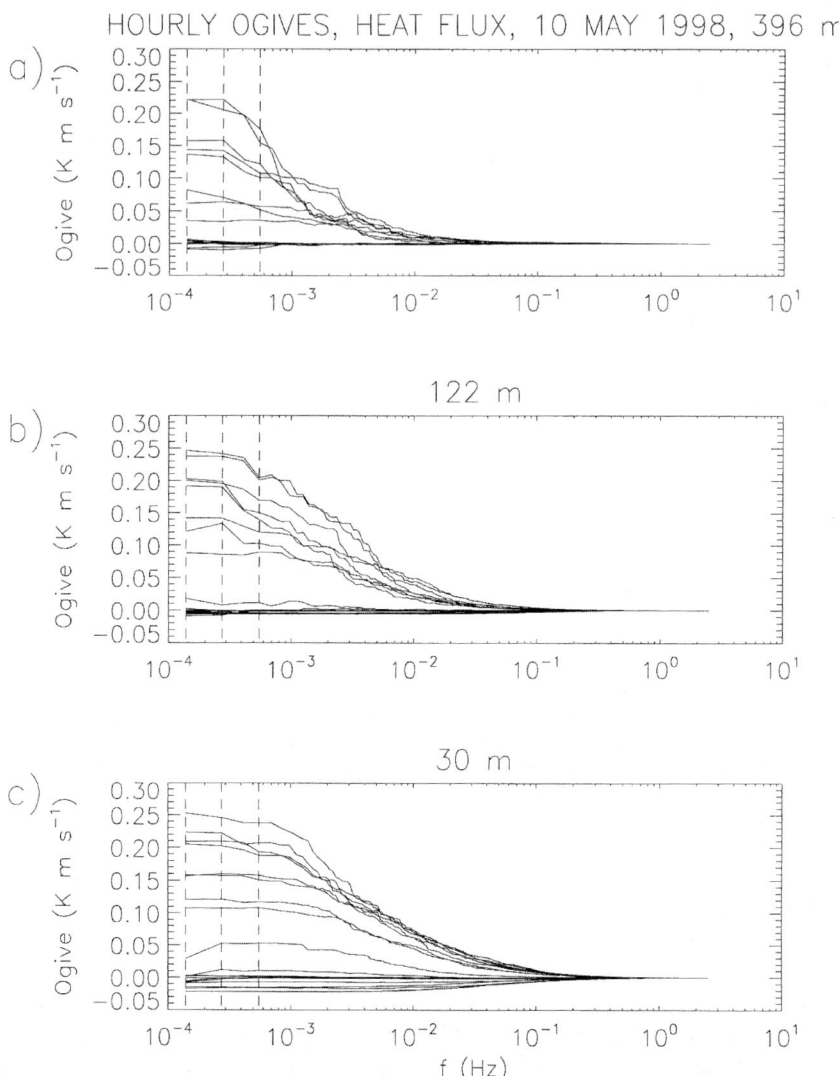

Figure 2.4. Ogives for the kinematic sensible heat flux at three measurement heights from the Chewamegon tall tower. The dotted vertical lines, from left to right, correspond to averaging periods of 120, 60 and 30 min respectively. Lines in each panel represent 2-hr sample periods centered at every hour of the day. Reproduced with permission of Berger et al. (2001).

ated with 'unwanted' non-stationarity might then contribute to the total covariance. Postponing for the moment the problem of dealing with deterministic non-stationary atmospheric motions, let us assume that we

Averaging, Detrending, and Filtering 21

have a background atmospheric state that is nearly steady over several hours. How do we determine if our system has sampled an adequate number of the larger eddies?

Traditionally, use has been made of empirical cospectral forms that reveal the contribution to the flux of eddies of different period. However, the best known of these 'standard spectra', those of Kaimal et al. (1972) were obtained over short vegetation surfaces and are not necessarily appropriate to use over tall forests. For example, Finnigan (2000) has pointed out systematic differences between the position of spectral and cospectral peaks in time series obtained over short vegetation surfaces and over tall canopies. A further cause for concern is that many standard spectra have been obtained using data processing techniques that have effectively high-pass filtered the contributions from low frequencies (Sakai et al. 2001, Finnigan et al. 2003).

An alternative is to use ogive plots that integrate under the cospectral curve and show the cumulative contribution of eddies of increasing period to the total transport (Figure 2.4). If the ogive curve reaches an asymptote at some period it indicates that there is no more flux beyond that period. In Figure 2.4, data from the tall tower site in Chewamegon (Berger et al. 2001) are used to compute ogives at different heights for different times of the day. They show that as measurement height increases, eddies of increasingly longer period make significant contributions to the total flux. To quantify this we have drawn three vertical dashed lines on the low frequency side. They correspond to sampling intervals of 30 min, 60 min and 120 min. It is clear that close to the ground (30 m) most of the ogives have reached an asymptote at averaging intervals of 30 min indicating that such an averaging interval was adequate at this height. At higher levels, however, the asymptote is not reached until around 60 min at 122 m and 120 min at 396 m. In other words, at 396 m an averaging time of two hours is necessary to capture all the flux.

The ogive and the cospectrum contain the same information because a point on the ogive plot is simply the integral under the spectral density curve between the highest frequency recorded and the frequency of interest. The advantage of the ogive presentation is that we can determine whether we have sampled for long enough by observing whether the ogive curve has reached its asymptote. We do not need to compare our measured spectra against some chosen standard that may be inappropriate for the measuring conditions we confront. The ogive is also useful when discussing the influence of low frequency fluxes on surface exchange (Chapter 5).

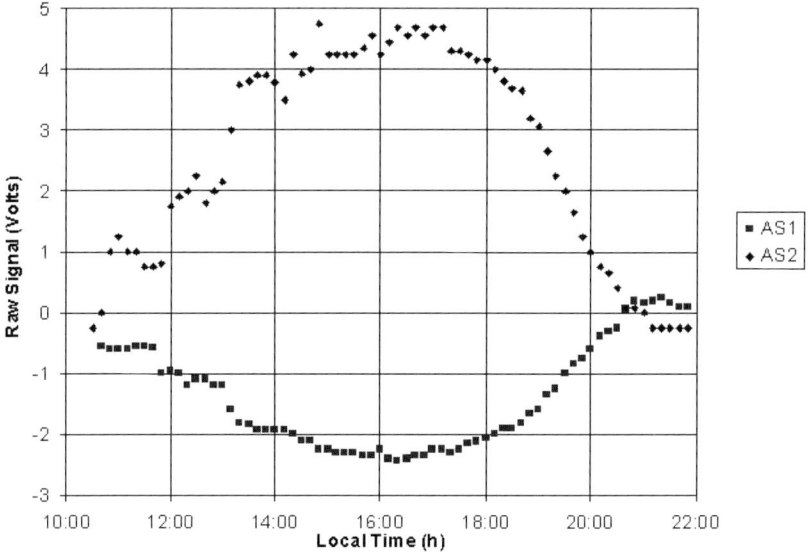

Figure 2.5. Short term drift in two outwardly identical open-path infra-red gas analyzers. (Moncrieff et al. 1992). Labels AS1 and AS2 refer to different instruments from the same manufacturer of open-path IRGAs.

4 The Origin of Low Frequency Content in the Signal

Trends[2] in data series can arise from two causes: instrumental drift and atmospheric changes. The latter include the advection of eddies of significant size over the measurement site, tropospheric processes like the passage of clouds that affect the surface energy balance, large scale changes of air mass and the evening and morning transitions in stability. All add low frequency content and non-stationarity to the data. The trend in the data can last from several minutes to many hours. The causes, and characteristics of low frequency atmospheric motions are discussed in Chapter 5. Here we discuss some of the instrumental causes.

To some extent the problem of instrumental drift is becoming less severe in that modern instruments such as infrared gas analyzers are now more stable in gain and offset. A dramatic example of how things have improved over the past decade or so is shown in Figure 2.5. Two

---

[2]For convenience we will use the term trend to refer to all low frequency components of our signal that we wish to separate from the true turbulent fluctuations. We have made a precise distinction, however between filtering and detrending.

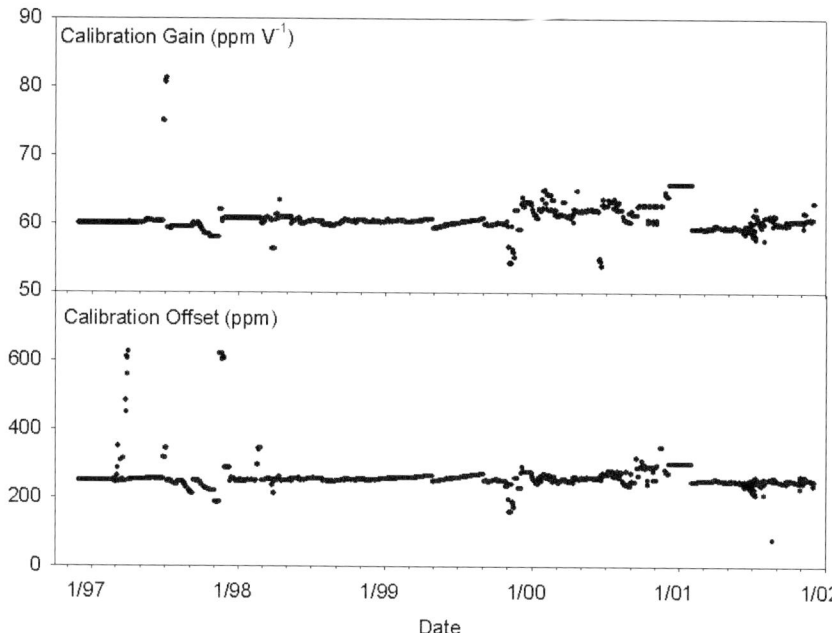

Figure 2.6. Long-term stability in gain and offset in a LI-COR 6262 at the Griffin field site, 1997-2001.

open-path sensors from the same manufacturer were co-located as part of an intercomparison during the FIFE experiment of 1989 (Moncrieff et al. 1992). Sensor 2 exhibited drift in the opposite direction to sensor 1 and twice the magnitude, even though they were set up identically. Under these circumstances, drift correction is real and had to be accounted for when producing fluxes using these instruments.

Compare that with the relative stability of a more modern closed-path IRGA (the LI-COR 6262) shown in Figure 2.6. In this figure, five years worth of span and drift records is shown for the period 1997-2001, obtained automatically at a flux site in Scotland (Clement et al. 2003). The relative constancy in gain over the whole period is clear although there are short periods of drift and change in offset. These changes can generally be explained with recourse to the field log, e. g. records of water getting into the sample tube and cell or changes to the method used to provide reference gas sample (changing from a chemical scrubber to a N_2 purge; changes to the software offset available with this instrument are also apparent). These records were obtained by the

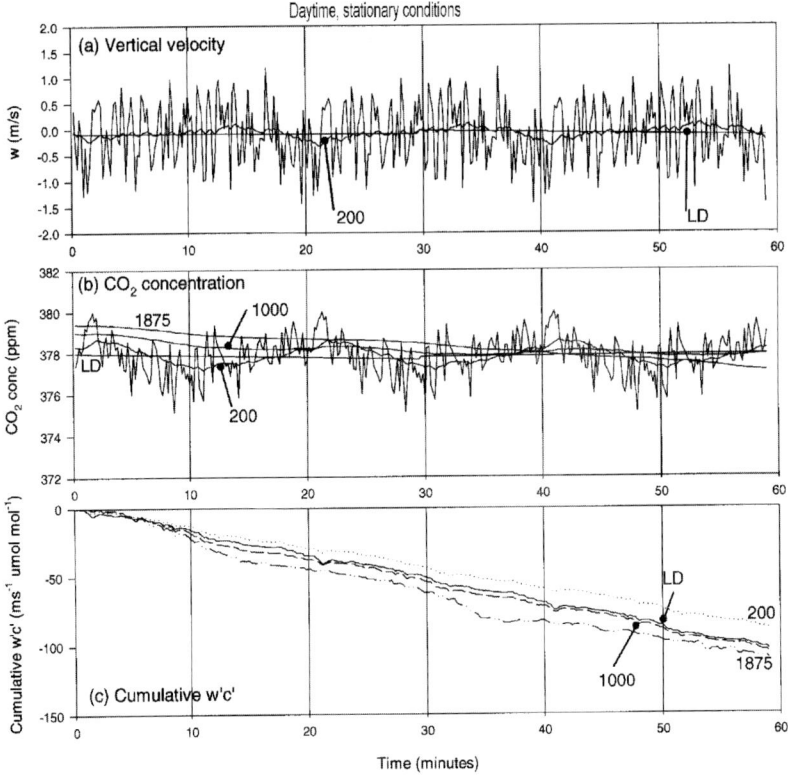

Figure 2.7. Midday turbulence data from two one-hour periods from the Manaus flux site of Kruijt et al. (2000) as illustrated by Culf (2000). Reproduced with permission of Culf (2000)

automatic calibration checking system which injects a scale and zero gas alternately into the sample cell every second day just after midnight.

Given this sort of instrument stability, detrending is less of an issue. One issue remains, however; what to do when the gain does drift over several days? Such long linear trends in data indicate problems with instruments since, in general, atmospheric phenomena do not exhibit linear trends in one direction for such periods. In this experiment, we argued that the effects were real and were amended in re-processing. It certainly points out the need to keep such records and good log books and is a good initial demonstration of data quality that is so important to the FLUXNET community.

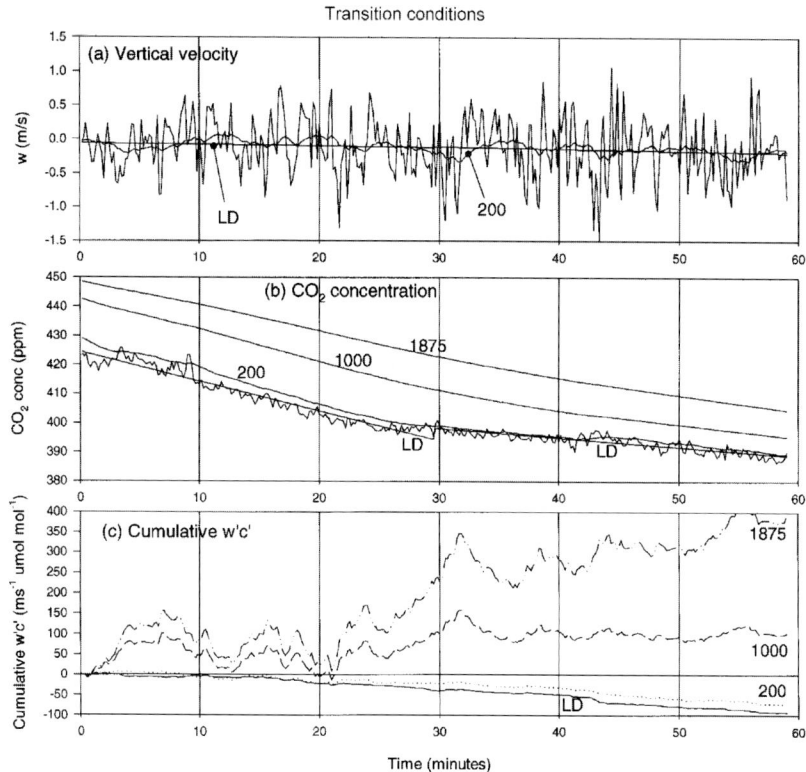

Figure 2.8. As in Figure 2.7 except for a transition case. Reproduced with permission of Culf (2000).

5 Errors Associated with Averaging, Filtering and Detrending

A thorough theoretical analysis of the effects of MR, LDT and filtering was performed by Lenschow et al. (1994) (henceforth, LMK). They assumed as a starting point two ideal signals $w(t)$ and $c(t)$. These signals were of infinite length, zero ensemble mean and had a joint co-correlation function that was exponential with an integral time constant τ. LMK were able to find analytic expressions for the differences between various statistics computed from the ideal series and the same statistics obtained after mean removal, linear detrending or filtering the series. They expressed these differences as errors and their results provide a benchmark for the application of these methods to real data that do not have the ideal characteristics assumed by LMK.

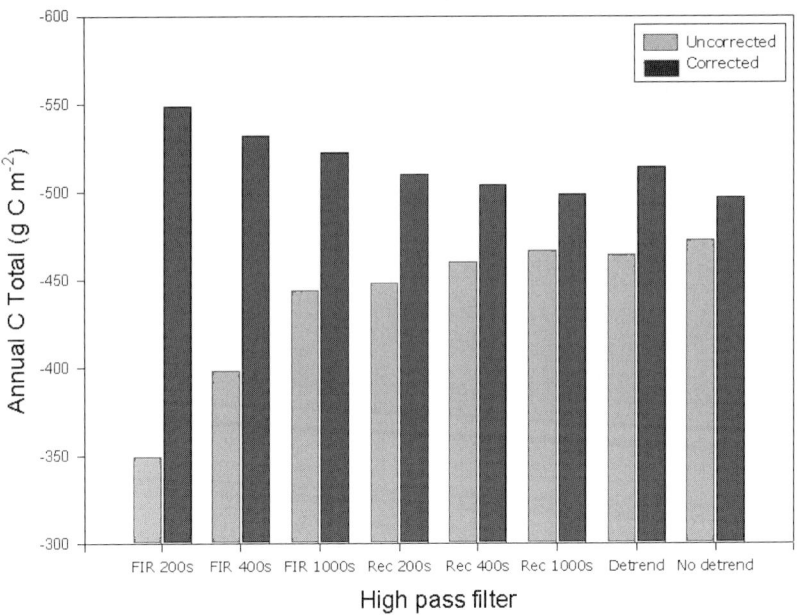

Figure 2.9. The influence of different detrending algorithms on data from one year at the Griffin forest field site in Scotland.

A more empirical example was provided by Culf (2000) and is shown in Figures 2.7 and 2.8. The figures show the influence of different detrending methods on short-term fluxes. Quite often we are interested in long-term fluxes (how much carbon is gained or lost by a forest over a year) but equally often we are interested to know how the vegetation is responding to environmental variables over the course of a few hours. Then, we need to examine much shorter intervals of time to calculate the fluxes. Here, Culf (2000) has taken 2 two-hour periods from a typical day in the Amazon at the Manaus site. The periods chosen are the middle of the day, representing stationary conditions, and a period of quite sharp transition in the early morning. During stationary conditions, the linear detrend and all three of the running mean filters produce similar fluxes over the period shown. In contrast, during the sharp morning transition at this site, whilst the linear detrend and 200 s running mean filter agree with each other, fluxes calculated by the longer running mean filters diverge markedly over the period shown.

Does the choice of detrending method matter over the course of a year? To answer this question, we re-processed flux data for 1998 from our forest site in Scotland with several different detrending methods. Figure 2.9 shows the difference between fluxes calculated with no detrend, a linear detrend, recursive filters of 200, 400 and 1000 s and Finite Impulse Response (FIR) filters of 200, 400 and 100 s. FIR filters have been popular in digital signal processing for some time given that they are easy to implement and do not distort the phase of the input signal (Holloway 1958, Ifeachor and Jervis 1993).

The frequency response correction to the data in Figure 2.9 followed the method of Moore (1986). The corrections were identical for block averaging (Kristensen (1998), sensor time constant mismatch (Massman et al. 1990), path-length averaging (Kristensen and Fitzjarrald 1984, Moore 1986), sensor separation (Kristensen 1979, Kristensen and Jensen 1979, Moore 1986), IRGA time constant (Moore 1986), and tube attenuation (Massman 1991). The corrections for signal detrending varied in accordance with the signal filtering method. The non-detrended fluxes had no correction applied for detrending. The frequency response transform of Aubinet et al. (2000) and Rannik and Vesala (1999) was applied to the linearly detrended data while a recursive filter frequency response transform (Moore 1986) was applied to the recursive-filtered data. The frequency response transform of the FIR filter was obtained by performing a Fourier transform on the filter weights.

Unsurprisingly, not making the correction for high-pass filtering produced underestimates of the annual total when compared to the no-detrend method. The short period FIR filters produced the largest underestimates and after correction, the greatest overestimate. The annual totals obtained by the recursive filters (all values) were within about 1% of the corrected value for no-detrend. The value for the linear detrend, corrected was within 3% of the corrected value obtained with no detrend. This is consistent with the results of Pekour et al. (2002). When this experiment was repeated but with an averaging period of 150 min, the annual carbon sequestered at this site increased by about 8%. The conclusion is a simple one: the choice of the most appropriate averaging period and detrending algorithm does matter if our estimates of carbon sequestration are to be as useful as we would wish to the rest of the community.

6 Averaging and Filtering in Complex Terrain: Special Considerations

Up until this point, our treatment has been conventional in that the analysis has assumed the underlying flow field is one dimensional. In complex topography, however, several other factors must be considered.

Most obvious is that the one-dimensional expression of the mass balance used in Section 2 must be replaced by the full three-dimensional form and steps taken to estimate horizontal advective and eddy flux divergences as well as the vertical flux, $\overline{w'c'}$. We must then confront the question of the appropriate period that separates 'deterministic' motions, that we will treat as advective fluxes, from stochastic turbulence, that we will treat as eddy fluxes. Although these questions are yet to be successfully resolved, they are discussed in Chapter 10 of this volume.

In complex terrain two methods are in use to rotate coordinates so that the fluxes can be analyzed in a framework with the z axis normal to the surface (or to the near surface streamlines) and the x and y axes in the plane tangent to the local surface. Until recently the usual method was to measure the components of the mean velocity vector, $\overline{\mathbf{u}} = \{\overline{u}, \overline{v}, \overline{w}\}$ in each averaging period T and then to rotate coordinates so that the x axis was parallel to $\overline{\mathbf{u}}$ and the cross-wind components, $\{\overline{v}, \overline{w}\} \to 0$ (e. g., Kaimal and Finnigan, 1994). It has now been shown (Finnigan et al. 2003) that this procedure has the effect of applying a high-pass filter to the velocity time series so that no fluctuations with period longer than T can contribute to the eddy flux. Furthermore, the transfer function of this implicit filter is rather complicated and the high frequency part of the cospectrum can be distorted because horizontal eddy flux $\overline{u'c'}$ and $\overline{v'c'}$ can be folded into $\overline{w'c'}$. There are other fundamental problems with the period-by-period rotation procedure and these are detailed in Finnigan (2004). In Chapter 3 of this volume it is recommended that the alternative 'planar fit' method be used to rotate coordinates in complex topography. The problems with implicit filtering when coordinates are rotated each period are not confined to complex terrain flows. If there is significant covariance in the wind and scalar fields at periods longer than the averaging/coordinate rotation period, T, then rotating coordinates each period T will remove it.

We have seen in Figures 2.3 and 2.4 that low frequency motions can make significant contributions to the eddy flux and we discussed some of the causes of this in Section 4. In complex topography, a further cause of long period 'eddies' can be added to the list. At such sites the average inclination of the wind vector to the vertical at the tower is often azimuth dependent. Finnigan et al. (2003) show an example of this from the Tumbarumba OzFlux site in New South Wales, Australia. In such cases, variations in wind direction translate into fluctuations of vertical wind speed at the tower. Azimuthal variations in the wind have periods from minutes to seasons and, if fluctuations in scalars are also azimuth dependent, they can combine to produce apparent low frequency content in the eddy flux. At the Tumbarumba site, energy balance closure

Averaging, Detrending, and Filtering 29

only occurs after averaging for 2 hours (Finnigan and Leuning, 2000). The cause of some of this low frequency content is almost certainly this mechanism.

7 Conclusions

- The purpose of averaging, detrending and filtering is to separate the active turbulent transport that we treat as eddy flux from slower, deterministic atmospheric motions and instrument drift.

- Three main tools are available to do this. The first two, averaging with mean removal and linear detrending are not filtering operations but their effect on the signal spectra can be approximated by transfer functions under certain conditions. Their advantages are simplicity and the fact that they can be applied to the full data record.

- True filtering is a convolution operation on the time series and requires a longer record than the section being filtered. The transfer function of a well-chosen filter window is cleaner than that of the mean removal or linear detrend, however, and the RC filter can be conveniently applied in a recursive way as data are gathered.

- In practice, the best method will be very dependent on conditions at a given site, including the data processing system being used. A comparison of alternative approaches is often wise.

- The period of fluctuations that are to be included in the eddy fluxes is generally longer at flux sites that measure continuously than micrometeorological data gathered under ideal conditions have led us to expect. Averaging times may have to be extended to as long as 4 hr to accommodate them. The origins of these long-period components of the eddy flux are various and include slow instabilities in the PBL, tropospheric forcing and, in complex topography, variations in vertical wind linked to azimuthal direction.

- To determine the optimal averaging period at any site, we have described an objective method that does not rely on assuming a co-spectral shape a priori. This so called 'ogive method' integrates the co-spectrum to successively longer periods until an asymptote is reached. The period of the asymptote can be regarded as the acceptable averaging period.

- No definitive method can be recommended to distinguish sensor drift from true low-frequency atmospheric signal although a linear trend maintained for very long periods should ring warning bells. A rigorous regime of sensor auto-calibration and recording of meta data is vital to recover true signal from data that is subsequently found to have been contaminated by sensor drift.

8 References

Aubinet, M., Grelle, A., Ibrom, A., Rannik, Ü., Moncrieff, J., Foken, T., Kowalski, A. S., Martin, P. H., Berbigier, P., Bernhofer, Ch., Clement, R., Elbers, J., Granier, A., Grünwald, T., Morgenstern, K., Pilegaard, K., Rebmann, C., Snijders, W., Valentini, R., Vesala, T.: 2000, 'Estimates of the annual net carbon and water exchange of European forests: the EUROFLUX methodology', *Adv. Ecol. Res.*, **30**, 113-175.

Bendat, J. S., Piersol, A. G.: 1958, *Measurement and Analysis of Random Data*, John Wiley and Sons, New York, 390 pp.

Berger, B. W., Davis, K. J., Yi, C., Bakwin, P. S. Zhao, C.: 2001, 'Long-term carbon dioxide fluxes from a very tall tower in a northern forest: Part I. Flux measurement methodology', *J. Atmos Oceanic Technol.*, **18**, 529-542.

Clement, R., Moncrieff, J. B., Jarvis, P. G.: 2003, 'Net carbon productivity of sitka spruce forest in Scotland', *Scottish Forestry*, **57**, 5-10.

Culf, A. D.: 2000, 'Examples of the effects of different averaging methods on carbon dioxide fluxes calculated using the eddy correlation method', *Hydrology & Earth System Sciences*, **4**, 193-198.

Finnigan, J. J.: 2000, 'Turbulence in plant canopies', *Ann. Rev. Fluid Mech.*, **32**, 519-571.

Finnigan, J. J.: 2004, 'A re-evaluation of long-term flux measurement techniques. 2: Coordinate systems', *Bound-Layer Meteorol.*, in press.

Finnigan, J. J., Leuning, R.: 2000 'Long term flux measurements — coordinate systems and averaging', In: *Proc. Int. Workshop Advanced Flux Network and Flux Evaluation*, 27-29 September 2000, Sapporo Japan, Center for Global Environmental Research, National Institute for Environmental Studies, Tsukuba, Japan.

Finnigan, J. J., Clement, R., Malhi, Y., Leuning, R., Cleugh, H.: 2003, 'A re-evaluation of long-term flux measurement techniques. 1: Averaging and coordinate rotation', *Bound.-Layer Meteorol.*, **107**, 1-48.

Gash, J. H. C., Culf, A. D.: 1996, 'Applying linear detrend to eddy correlation data in real time', *Bound.-Layer Meteorol.*, **79**, 301-306.

Holloway, J. L.: 1958, 'Smoothing and filtering of time series and space fields', *Adv. Geophys.*, **4**, 351-388.

Ifeachor, E. C., Jervis, B. W.: 1993, *Digital Signal Processing: A Practical Approach*, Addison-Wesley, Harlow, 760 pp.

Kaimal, J. C., Wyngaard, J. C., Izumi, Y., Cote, O. R.: 1972, 'Spectral characteristics of surface layer turbulence', *Quart. J. R. Meteorol. Soc.*, **98**, 563-589.

Kaimal, J. C., Finnigan, J.: 1994, *Atmospheric Boundary Layer Flows: Their Structure and Measurement*, Oxford University Press, Oxford.

Kristensen, L.: 1998 *Time Series Analysis: Dealing with Imperfect Data*, Riso National Laboratory, Roskilde, Denmark, Riso-I-12289(EN) pp31.

Kristensen, L., Fitzjarrald, D.: 1984, 'The effect of line averaging on scalar flux measurement with a sonic anemometer near the surface', *J. Atmos. Oceanic Technol.*, **1**, 138-146.

Kristensen, L.: 1979, 'On longitudinal spectral coherence', *Bound.-Layer Meteorol.*, **16**, 145-153.

Kristensen, L., Jensen, N. O.: 1979, 'Lateral coherence in isotropic turbulence and in the natural wind', *Bound.-Layer Meteorol.*, **17**, 353-373.

Kruijt, B., Malhi, Y., Lloyd, J., Miranda, A. C., Nobre, A. D., Pereira, M. G. P., Culf, A., Grace, J.: 2000, 'Turbulence above and within two Amazon rainforest canopies', *Bound.-Layer Meteorol.*, **94**, 297-331.

Lenschow, D. H., Mann, J., Christens, L.: 1994, 'How long is long enough when measuring fluxes and other turbulence statistics?', *J. Atmos. Oceanic Technol.*, **11**, 661-673.

Massman, W. J.: 1991, 'The attenuation of concentration fluctuations in turbulent flow through a tube", *J. Gephys. Res.*, **96**, 15269-15273.

Massman, W. J., Fox, D. G., Zeller, K. F., Lukens, D.: 1990, *Verifying Eddy Correlation Measurements of Dry Deposition: A Study of the Energy Balance Components of the Pawnee Grasslands*, USDA Forest Service Research Paper RM–288, Rocky Mountain Forest and Range Experiment Station, Fort Collins, CO, 14 pp.

McMillen, R. T.: 1988, 'An eddy correlation technique with extended applicability to non simple terrain', *Bound.-Layer Meteorol.*, **43**, 231-245.

Moncrieff, J. B., Verma, S. B., Cook, D. R.: 1992, 'Intercomparison of eddy correlation carbon dioxide sensors during FIFE 1989', *J. Geophys. Res.*, **97**, 18725-18730.

Moore, C. J.: 1986, 'Frequency response corrections for eddy correlation systems', *Bound.-Layer Meteorol.*, **37**, 17-35.

Pekour, M. S., Wesely, M. L., Martin, T. J., Cook, D. R.: 2002, 'A study of block averaging versus recursive filters for computing scalar eddy covariances near the surface', In: *American Meteorol. Soc. 25th Conf. Agric. Forest Meteorol.*, Norfolk, VA, p144-45.

Rannik, Ü., Vesala, T.: 1999, 'Autoregressive filtering versus linear detrending in estimation of fluxes by the eddy covariance method', *Bound.-Layer Meteorol.*, **91**, 259-280.

Sakai, R. K., Fitzjarrald, D. R., Moore, K. E.: 2001, 'Importance of low-frequency contributions to eddy fluxes observed over rough surfaces', *J. Appl. Meteorol.*, **40**, 2178-2192.

Chapter 3

COORDINATE SYSTEMS AND FLUX BIAS ERROR

Xuhui Lee, John Finnigan, Kyaw Tha Paw U
xuhui.lee@yale.edu

Abstract   This Chapter examines theoretical and operational aspects of coordinate systems. A distinction is made between the vector basis, a local property of a coordinate system, and the overall coordinate frame consisting of the vector basis and coordinate lines, a global property of the flow that is determined by the flow field in three dimensions. Point measurements can only define the vector basis. Because in field campaigns many components that enter into the mass balance in complex flows are severely under-sampled, a properly chosen coordinate frame for point measurements should optimize our estimates of the surface-air exchange and should maximize information for diagnostics purposes.

The strengths and weaknesses of three operational coordinate systems for point measurements (instrument, natural wind, and planar fit) are examined in detail. That error in scalar fluxes due to coordinate tilt is usually small for small tilt angles does not negate the need for coordinate rotation because the tilt error can introduce a systematic bias to the time integrated flux. On the other hand, it is also important that over-rotation be avoided in post-field data analysis. Tilt errors caused by contamination from the streamwise and cross-wind fluxes should be treated differently.

Appendix B outlines a method for rotation into the planar fit coordinate. The scheme relies on the straightforward vector operation and avoids the need for rotation angles.

1   Introduction

Application of coordinate rotation is a necessary step in micrometeorological studies of surface-air exchange before the observed fluxes can be meaningfully interpreted. The most common rotation procedure uses measured mean wind to define an orthogonal vector basis, termed natural wind system, for each observational period (e. g., 30 min) to which all fluxes are transformed. The rotation scheme is intended to level the sonic anemometer to the terrain surface. When it was first proposed by Tanner and Thurtell in 1969 [see also Kaimal and Finnigan (1994) and McMillen (1988)], the natural wind system was limited to a surface layer in which the flow is one dimensional, that is, the velocity and scalar concentration gradients exist only in the vertical and hence no horizontal scalar advection nor flow divergence, and there is no wind directional shear causing a cross-wind momentum flux. It appeared sufficient from the 1960's through the early 1990's as most field experiments then were conducted at ideal sites, over selected "golden days", and in fair weather conditions. The scope of micrometeorological research has now been extended considerably, to include non-ideal sites and year-round, continuous monitoring, and the validity of the procedure is now called into question.

More recent rotation schemes (Wilczak et al. 2001, Paw U et al. 2000, Lee 1998) attempt to overcome some of the deficiencies of the natural wind system. However, like Tanner and Thurtell (1969), they do not in fact treat coordinate systems at all but focus rather on the orientation of the vector basis, $\vec{e}_i$, in which vector and tensor quantities are to be represented. This is an important and continuing question as the circumstances of most flux sites dictate that the wind field itself must be used to orient $\vec{e}_i$. The vector basis is a local property of a coordinate system but it is the global properties of the flow field that dictate the form of the mass balance equation that we employ to convert flux measurements to measures of surface exchange and so it is vital to understand the relationship between the two quantities as well as the advantages and disadvantages of different coordinate systems.

Removal of "tilt errors" or cross-contamination among components of the eddy flux vector is cited in the literature as the main reason for performing coordinate rotation. Kaimal and Haugen (1969) and others have shown that momentum flux is particularly sensitive to the tilt errors. Scalar fluxes are not as sensitive, but the errors could potentially cause a systematic bias in annually integrated eddy fluxes (Section 4). It is known that a tilt-corrected flux does not necessarily represent the true surface-air exchange because non-turbulent advective components of the surface-layer mass balance may be non-negligible even at ideal sites. A proper coordinate frame is vitally important to advance our understanding of these issues.

Strictly, to use measurements of wind speed, concentration and eddy flux to infer surface exchange of a scalar c involves the assimilation of measurements into a description of the mass balance in a control volume V, erected over a representative patch of the surface (e. g., Figure 6.1 of Chapter 6). The mass balance of c is the sum of the fluxes of c across each face of the control volume plus the accumulation of c within the volume. If we can measure the fluxes across each aerial face as well as the rate of change of c within V, we can deduce the transfer across the surface by difference. Whatever kind of instrumentation we employ, however, we are only able to sample the aerodynamic flux and the rate of change of c at a few points in space and we are forced to either supply the missing information in other ways or develop good diagnostic tools that can aid selective use of data.

The mathematical form of the mass balance that we employ has a considerable bearing upon our ability to estimate its constituent terms from a finite number of measurements. The two main factors affecting this form are the averaging operations applied to the instantaneous variables and the coordinate system in which the mass balance is represented. The question of averaging operators and their relationship to coordinate alignment is dealt with in detail in Finnigan et al. (2003) and Sakai et al. (2001) although there, the only coordinate system considered is the familiar rectangular Cartesian frame. Here we concentrate on the choice of coordinate system and assume that an appropriate averaging operator may be applied to the measurements.

This Chapter examines theoretical and operational aspects of coordinate systems. It begins with a brief discussion of the theoretical constraints on the coordinate system. [The reader is referred to Finnigan (2004) for more details.] This is followed by a discussion on the strengths and weaknesses of three common coordinate frames for point measurements, the instrument coordinate, the natural wind system, and the planar fit coordinate (Section 3). Section 4 provides a assessment of flux bias errors due to sensor tilt in horizontally homogeneous flow. Section 5 discusses examples of coordinate tilt that are likely to occur in field observations. In Section 6, a dataset obtained over a forest in complex terrain is analyzed to examine the sensitivity of flux calculation to coordinate rotation.

2 Theory

2.1 Mass balance at a point

A coordinate frame is meaningful only if it is consistent with the frame used by equations that comprise, either explicitly or implicitly, the theory underlying the study. Kaimal and Finnigan (1994) state, "... problems occur when vector quantities like velocities or fluxes are measured in a reference framework that does not coincide with that of the equations used to analyze them". The fundamental equation for surface-air exchange studies is the mass conservation equation. Although in a strictly formal analysis, all fluxes can be expressed as 3-dimensional vectors with the gradient operator being independent of coordinate system (Massman and Lee 2002), in practice a coordinate frame is needed to estimate the surface-atmosphere exchange and the related turbulent statistics, including the net ecosystem exchange (NEE).

The statement of conservation of a scalar c at a point in an incompressible fluid is

$$\frac{\partial c}{\partial t} + \nabla . \vec{u}c = S(\vec{x})\delta(\vec{x} - \vec{x}_0) \qquad (3.1)$$

where the velocity vector $\vec{u}$ has components u, v, w corresponding to position vector $\vec{x}$ with components x, y, z. The source term S is multiplied by the Dirac delta function, signifying that the source is zero except on the ground and vegetation surfaces, whose locus is $\vec{x}_0$. We have ignored molecular diffusion, which is negligible except very close to solid surfaces when its effects can be conveniently absorbed in the specification of the source strength, for example via the device of a boundary-layer resistance. The scalar c represents any absolute fluid property such as density of carbon dioxide or heat content. For alternative formulations of the mass balance see Paw U et al. (2000) and Raupach (2001).

Each term in Equation 3.1 is a scalar and so is independent of the coordinate frame. The individual components of the divergence term, however, take different forms in different coordinate systems. There are three overriding requirements guiding the choice of coordinate frame and its orientation

- We must be able to express our measurements in the chosen coordinate frame.

- Since we can rarely measure all the components of $\nabla . \vec{u}c$, we want to work in a coordinate frame that optimizes our ability to estimate $\nabla . \vec{u}c$, using the terms we can measure.

- If we want to assimilate our measurements explicitly into a mathematical model of flow and transport, we would like to be able to construct such a model in the chosen coordinates.

In this Chapter we consider only the first two of these requirements.

We can illustrate the dependence of the form of the flux divergence upon coordinate frame and orientation most simply through the example of one-dimensional flow over horizontally homogeneous terrain. In this case an appropriate coordinate system is the rectangular Cartesian

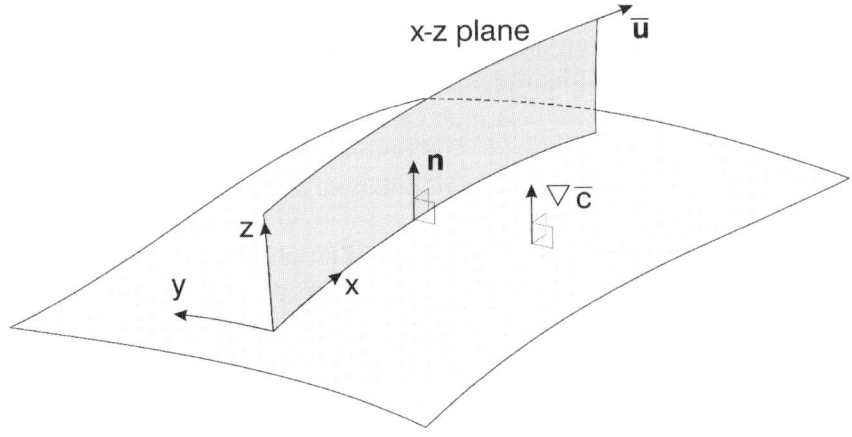

Figure 3.1. The coordinate system should be such that the local normal to the surface and the mean scalar gradient $\nabla \bar{c}$ lie in the x-z plane.

frame and with an arbitrary orientation of the axes and with velocity components u, v, w aligned with x, y, z respectively, the divergence of the mean aerodynamic flux vector becomes,

$$\nabla . \overline{\vec{u} c} = \frac{\partial \overline{uc}}{\partial x} + \frac{\partial \overline{vc}}{\partial y} + \frac{\partial \overline{wc}}{\partial z} \quad (3.2)$$

where the overbar denotes a time average.

One-dimensionality of the wind field and horizontal homogeneity of the surface scalar source impose strong symmetries on the velocity and scalar fields so that gradients of mean quantities depend only on distance from the surface. Hence, if we orient $\vec{e}_i$ so that the z axis is normal to the surface we find

$$\nabla . \overline{\vec{u} c} = 0 + 0 + \frac{\partial \overline{wc}}{\partial z} \quad (3.3)$$

In this case the divergence operator can be estimated, at least in finite difference form, from anemometers and scalar sensors on a single tower orientated along the surface-normal z axis. A more general message can be drawn from this example, however. It reminds us that the major symmetries of the wind field and the scalar source distribution determine the alignment of gradients of mean moments of the velocity and scalar fields. In fact it is the symmetry of the wind field that the natural wind system was built upon.

If we move from one-dimensional to two- and three-dimensional flows, we expect that the alignment of flow streamlines[1], which will now be space curves, and the directions in which the scalar source distribution changes most rapidly will continue to determine the strongest symmetries of the resultant mean fields and thereby the gradient of the aerodynamic flux vector. Consider for example, boundary-layer flow over gently undulating terrain with horizontal changes in scalar source strength on scales of kilometers or greater. The mean streamlines close to the surface will be approximately parallel to the ground while the gradients of mean moments of the wind and scalar fields in the surface-normal, cross-streamline direction will be much larger than streamwise gradients. Hence we can write

$$\frac{\partial \overline{wc}}{\partial z} >> \frac{\partial \overline{uc}}{\partial x}, \frac{\partial \overline{vc}}{\partial y} \qquad (3.4)$$

where the x and y directions are now aligned in the streamwise direction and in the cross-stream direction parallel to the local surface, respectively. Equivalently we can say that the local normal to the surface must lie in the x-z plane (Figure 3.1). In analogy to the case of one-dimensional flows, the best approximation to the divergence that can be obtained from an alignment of anemometers along a single tower is obtained when the instruments are located in the plane spanned by the mean wind vector and the local normal to the surface.

Flows where the mean streamlines are approximately parallel to the surface and Equation 3.4 is satisfied are sometimes referred to as 'Fairly Thin Shear Layers' (FTSL) (Bradshaw, 1973). Most long-term flux study sites conform to the FTSL description, even those in complex terrain. Henceforth, we will refer to terrain where the flow satisfies FTSL criteria as 'gentle' terrain. At such locations, we can expect that measuring $\overline{wc}/\partial z$ with x tangent to the streamline and the x-z plane normal to the surface will yield the best approximation to $\nabla . \vec{uc}$ that we can obtain from instruments orientated along a single straight line. For a practical example of this see Geissbuhler et al. (2000). As the scale of variation of the mean velocity and the scalar source in the streamwise direction begins to approach that in the cross-stream direction, however,

---

[1]Streamlines are curves in space that are everywhere tangent to the local velocity vector. The streamlines passing through an arbitrary curve that is not itself a streamline form a stream surface. If the velocity vector $\vec{u}$ is a time averaged quantity, then the streamlines and stream surfaces belonging to the steady vector field $\vec{u}(\vec{x})$ are fixed in space. Solid surfaces are stream surfaces by definition, as the normal component of $\vec{u}$ is zero at such a surface. For more complete definitions of these objects see any standard text on fluid mechanics, e. g., Batchelor (1967) and for a comment on the limitations of the concept of a stream surface see Finnigan (1990).

this approximation rapidly becomes poor. Nevertheless, at micrometeorological sites chosen to avoid the grossest inhomogeneities in topography and source distribution, the optimal coordinate system in which to write the mass balance is one whose coordinate lines are aligned as shown in Figure 3.1.

So far we have concentrated on the mass balance at a point and the *local* orientation of the coordinate lines. In practice we want to estimate the mass balance in a control volume over a representative patch of surface. To do this we need to write the mass balance in integral form, which requires us to specify the coordinate system in which we intend to represent it as this determines the geometry of the coordinate lines along which we shall integrate. In the next section we will review the properties of two candidate systems whose coordinate lines have the *local* orientation specified above.

2.2   Coordinate systems

Coordinate systems provide two essential ingredients for the mathematical description of the mass balance: they specify the magnitude and direction of a vector basis $\vec{e}_i$ in terms of which all vector and tensor quantities can be written, e. g.

$$\vec{u} = u\vec{e}_1 + v\vec{e}_2 + w\vec{e}_3 \tag{3.5}$$

u, v, w being the components of the velocity vector $\vec{u}$ in the basis $\vec{e}_i$. They also provide coordinate lines, whose intersections can be used to locate points in space and along which we integrate, e. g., we write $\vec{u}(\vec{x}) \equiv \vec{u}(x, y, z)$ meaning the value of vector $\vec{u}$ at the position labeled by distances x, y, z, respectively from the origins of the coordinate lines. The vector basis, $\vec{e}_i$ is linked to the coordinate lines. For example, $\vec{e}_1$ might be defined as the unit tangent to the x coordinate line.

Except in the simplest case of steady one-dimensional flow over a plane surface, in which case the mean streamlines are straight lines parallel to the surface, a coordinate system that has its x lines approximately parallel to and its z lines normal to the streamlines will be curvilinear. Some salient points of curvilinear coordinate systems together with some useful references are given by Finnigan (2004). In the next section we will discuss two coordinate systems in detail: rectangular Cartesian and physical streamline coordinates, which essentially bound the range of appropriate choices.

2.2.1 Rectangular Cartesian coordinates

In this familiar system the vector basis $\vec{e}_i$ is orthonormal and the coordinate lines are straight and orthogonal and everywhere parallel to $\vec{e}_i$ so that the x coordinate is parallel to $\vec{e}_1$, y is parallel to $\vec{e}_2$ and z is parallel to $\vec{e}_3$. The instantaneous mass balance equation (Equation 3.1) written in Cartesian coordinates (from now on we will drop the qualification 'rectangular') is

$$\frac{\partial c}{\partial t} + \frac{\partial uc}{\partial x} + \frac{\partial vc}{\partial y} + \frac{\partial wc}{\partial z} = S\delta(\vec{x} - \vec{x}_0) \qquad (3.6)$$

and the time averaged form of this equation is

$$\frac{\overline{\partial c}}{\partial t} + \overline{u}\frac{\partial \overline{c}}{\partial x} + \overline{v}\frac{\partial \overline{c}}{\partial y} + \overline{w}\frac{\partial \overline{c}}{\partial z} + \frac{\partial \overline{u'c'}}{\partial x} + \frac{\partial \overline{v'c'}}{\partial y} + \frac{\partial \overline{w'c'}}{\partial z} = \overline{S}\delta(\vec{x} - \vec{x}_0) \quad (3.7)$$

where the overbar denotes a simple time average (Finnigan et al. 2003) and the prime denotes an instantaneous departure from the average.

An important property of rectangular Cartesian coordinates is that, once the vector basis has been defined at any point in space, its orientation and that of the coordinate lines is defined everywhere (Figure 3.2). In particular, if we determine the x, $\vec{e}_1$ direction by making it parallel to the mean velocity vector measured by a sonic anemometer on a tower and if the mean streamline at the anemometer is not parallel to the underlying surface, then the z axis cannot be normal to the surface.

2.2.2 Physical streamline coordinates

Physical streamline coordinates are defined by the flow field itself. The instantaneous flow must first be averaged in time to define a set of mean streamlines, which become the x coordinate lines. Hence a given turbulent flow field can generate different streamline coordinate frames depending upon the way the flow is averaged. Like Cartesian coordinates, streamline coordinates employ the orthonormal basis $\vec{e}_i$ but this is now orientated so that $\vec{e}_1$ is always tangent to the local streamline, $\vec{e}_2$ is aligned with the *principal normal*[2] to the streamline and $\vec{e}_3$ is aligned

---

[2] The principal normal to a streamline lies in the plane that is tangent to the streamline and in which the curvature of the streamline is greatest. The binormal is perpendicular to the plane spanned by the tangent and the principal normal and the three vectors, the tangent, principal normal and binormal form the orthonormal *Frenet Frame*. In two-dimensional flow fields the binormals and the y coordinate lines are parallel to the surface so that the z coordinate lines intersect the surface normally. Hence the physical streamline coordinates of horizontally homogeneous flow over a flat surface are just rectangular Cartesian coordinates with the z axis normal to the surface.

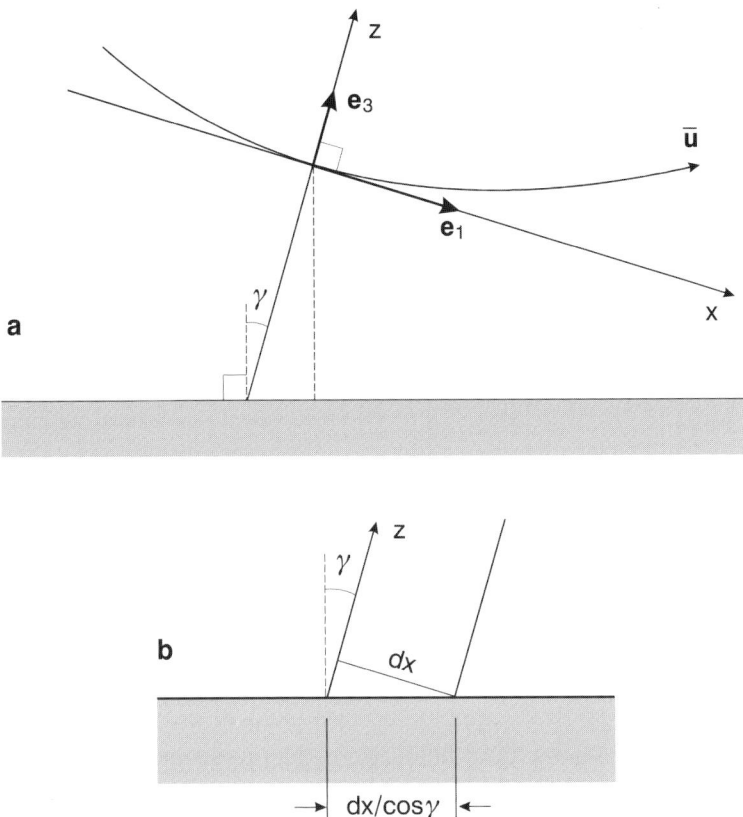

Figure 3.2. a. The orientation of a Cartesian coordinate system is determined once the base vectors are orientated at a single point, usually the anemometer position. b. If the z axis intersects the ground at an angle γ, the area of ground surface that supplies a flux of mass into the prism $\mathrm{d}x \times \mathrm{d}y \times z$ is $\mathrm{d}x \times \mathrm{d}y / \cos\gamma$.

with the binormal to the streamline (Figure 3.3). The coordinate lines x, y, z are respectively, the streamlines (x), the set of curves everywhere tangent to the binormals (y) and the set of curves everywhere parallel to the principal normals (z) (Figure 3.3). Note that in physical streamline coordinates, the y coordinate lines are associated with the $\vec{e}_3$ base vectors and the z lines with $\vec{e}_2$. This is a consequence of the micrometeorological convention where we take the positive z direction as increasing normally from the surface. Streamline coordinates are described for two-dimensional flows by Finnigan (1983) and for three-dimensional flows by Finnigan (1990) and Kaimal and Finnigan (1994). Their application to long term flux measurements is treated in much more detail in Finnigan (2004). Two-dimensional streamline coordinates have been employed

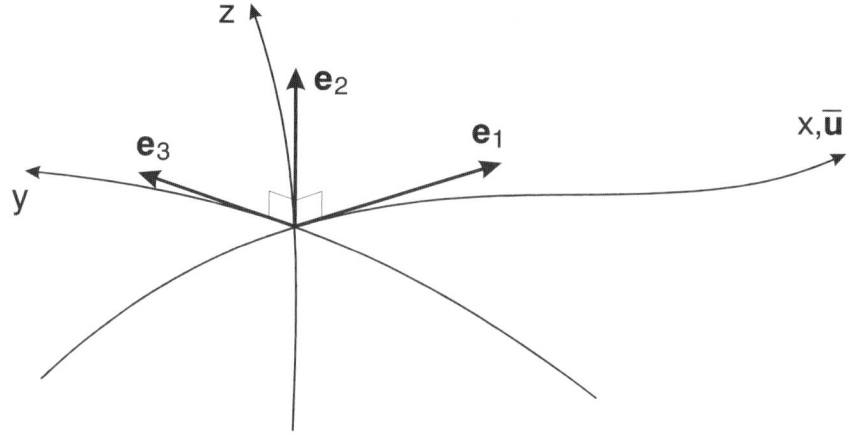

Figure 3.3. The vector basis of physical streamline coordinates is determined by the orientation of the streamline. Here $\vec{e}_1$ is the (normalized) tangent, $\vec{e}_2$ the (normalized) principal normal and $\vec{e}_3$ the (normalized) binormal to the streamline. The z coordinate lines are tangent to the field of $\vec{e}_2$ vectors and the y coordinate lines are tangent to the field of $\vec{e}_3$ vectors, while the streamlines form the x coordinates.

in analyses of complex flow fields, e. g. Finnigan and Bradley (1983), Zeman and Jensen (1984).

The time averaged mass conservation equation in three-dimensional streamline coordinates is

$$\frac{\partial \bar{c}}{\partial t} + \bar{u}\partial_x \bar{c} = -\partial_x \overline{u'c'} - \partial_y \overline{v'c'} - \partial_z \overline{w'c'} - [\frac{1}{L_a}]\overline{u'c'} + [\frac{1}{r}\frac{\partial r}{\partial y}]\overline{v'c'}$$
$$-[\frac{1}{R} + \frac{1}{r}]\overline{w'c'} + S\delta(\vec{x} - \vec{x}_0) \quad (3.8)$$

where $\dfrac{1}{L_a} = \dfrac{1}{\bar{u}}\partial_x \bar{u}$, R is the local radius of curvature of the streamline and r is the local radius of curvature of the y coordinate lines. One consequence of using curvilinear systems like streamline coordinates but retaining the orthonormal vector basis $\vec{e}_i$ so variables have their familiar meaning is that the derivatives in the equations are directional rather than partial derivatives and we have written them as ∂_x, ∂_y etc. to distinguish them from partial derivatives. However, for most practical applications directional and partial derivatives are interchangeable. The main difference that needs to be kept in mind when doing mathematical manipulation of streamline coordinate equations is that derivatives along orthogonal coordinate lines do not commute so $\partial_x \partial_y \phi - \partial_y \partial_x \phi \neq 0$, where $\phi(x, y, z)$ is an arbitrary function (Finnigan, 1990). Momentum equations and rate equations for the components of the Reynolds stresses in this coordinate frame may be found in Kaimal and Finnigan (1994).

We see in Equation 3.8 that in streamline coordinates the advection term has been simplified relative to Equation 3.7 so that only streamwise advection appears in the equation but that the flux divergence has acquired extra terms that arise because of the changing orientation of $\vec{e}_i$ in space and because the infinitesimal control volume $dxdydz$ changes shape as streamlines converge or diverge. On comparing the eddy flux terms in Equations 3.7 and 3.8 it is apparent that these extra terms all involve the radii of curvature of the coordinate lines.

2.2.3 Other coordinate systems

Mathematical models of flow and transport over complex terrain are often developed in various kinds of surface-following coordinate systems. See for example, Howarth (1951), Bradshaw (1973), Pielke (1984), Ferziger and Peric (1997). While these systems offer advantages for constructing models, they have significant disadvantages for interpreting tower measurements. The main ones are that the coordinate systems are generally non-orthogonal and the associated vector bases are not orthogonal unit vectors so that the dependent variables in these systems do not correspond to the physical quantities that our instruments measure. We will not discuss such systems further here. For a more detailed appreciation see Finnigan (2004).

2.3 Advantages and disadvantages of Cartesian and streamline coordinate systems

A comparison of Equations 3.7 and 3.8 shows that the mass balance expressed in streamline coordinates has a simplified advection term but a more complicated expression for the flux divergence with three extra terms to estimate. In gentle terrain it is probably easier to estimate the parameters L_a, R and r than it is to estimate the cross stream advection terms $\overline{v}\partial\overline{c}/\partial y + \overline{w}\partial\overline{c}/\partial z$ as R and r can be approximated as $R = R_0 + z$ and $r = r_0 + z$, where R_0 and r_0 are the curvatures of the surface and can be calculated from a digital elevation or contour map. In steeper topography, however, this advantage is lost and multipoint measurements are required to close the mass balance whichever coordinate frame it is written in.

Another advantage of streamline coordinates is that the vector basis $\vec{e}_i$ is everywhere aligned with the local mean wind vector $\vec{\overline{u}}$ so that a series of anemometers can be combined in the mass balance calculation once their outputs have been rotated into the local $\vec{e}_i$ basis. In Cartesian coordinates, in contrast, once the orientation of $\vec{e}_i$ has been determined for one anemometer, it is fixed everywhere in space and the orientations of additional anemometers relative to the first must be known to use their outputs in the mass balance calculation.

To move from the time averaged mass balance at a point as expressed by Equations 3.7 and 3.8 to the integral mass balance in a control volume, we need to integrate Equations 3.7 and 3.8 over a prism whose lower face is the vegetated surface and whose aerial faces are determined by the coordinate surfaces. This raises the issue of determining these surfaces which in gentle terrain is somewhat easier in the case of streamline coordinates than in Cartesian coordinates. In steep terrain however, neither system is obviously superior to the other. A more complete treatment of the pros and cons of the two coordinate systems may be found in Finnigan (2004).

3 Coordinate Systems for Point Measurements

3.1 General considerations

This Section discusses three coordinate frames that are used most frequently for the interpretation of point eddy covariance measurements. These coordinate frames all define a local vector basis in which vector quantities such as air velocity and eddy flux are expressed. In addition, none of them uses the scalar concentration and flux fields to constrain the vector basis. This second feature is an important one because any other coordinate systems constrained in whole or in part by the scalar flux vector will give physically unrealistic results.

In Section 2, we make a distinction between the vector basis, a local property of a coordinate system, and the overall coordinate frame consisting of the vector basis and coordinate lines, a global property of the flow that is determined by the flow field in three dimensions. From an operational viewpoint, point measurements can only define the local vector basis. Even with multiple sensors, it is extremely difficult to determine coordinate lines of the global coordinate frame because the sensors can rarely be aligned relative to one another with sufficient accuracy. Furthermore, point measurements give some but not all of the terms of the surface-layer mass balance. It is therefore crucial that we work in coordinate frames that optimizes our ability to estimate surface-vegetation exchange such as NEE, using the terms we can measure. A suitable coordinate frame must also maximize information for diagnostics purposes (e. g., to answer the question of whether atmospheric conditions are too limiting to allow a meaningful NEE estimate) and for advancing our understanding of the 3-dimensional nature of the flow.

Unfortunately, in the authors' opinion, the information produced by eddy covariance has not been fully utilized because most field studies focus too narrowly on the vertical eddy fluxes such as CO_2 flux and the streamwise momentum flux. It is known that even at ideal sites, the 30-min mean velocity vector can depart from the local terrain surface. Recovery of the mean vertical velocity may help us determine whether the observation suffers from undersampling of low frequency eddies or from the influence of mesoscale motion at scales larger than the scale of the flux footprint. It is also known that $\overline{v'w'}$, the cross-wind momentum flux, cannot be assumed to equal zero in the ocean atmospheric surface layer (Wilczak et al. 2001), at sites on rolling topography (Section 6), and at times when wind directional shear exists in the surface layer. Tanner and Thurtell (1969) pointed out that when $\overline{u'v'}$ (covariance between the streamwise and lateral velocity components) is not zero, conditions are not ideal and local divergence caused by fetch or surface inhomogeneity may be occurring. Lee (2004) discussed the mechanism of generation of the horizontal eddy flux, $\overline{u'c'}$, in the surface layer and how it can be used provide additional information on the advective influences on flux observations. These quantities are physically meaningful only if a coordinate is chosen properly.

A suitable coordinate system also provides a consistent framework for data analysis. This is especially true if one wishes to recover flux loss at low frequencies (Finnigan et al. 2003, Sakai et al. 2001, Chapter 5). In this regard, rotation at every 30 min interval, which is equivalent to high-pass filtering, produces the undesirable effect of having turbulent time series that are discontinuous. Similarly, construction of ensemble mean spectra and cospectra should be done in an appropriate coordinate frame so that the low frequency contributions to the spectra are not missed.

3.2 Instrument coordinate

This is an orthogonal coordinate frame deployed by the anemometer to express the components of the wind and the associated eddy flux vectors. In some modern designs, the transducers of the sonic anemometer are arranged non-orthogonally to minimize flow interference. Projection of the velocity vector from the non-orthogonal to the desired orthogonal frame involves straightforward geometric transformation, which is ac-

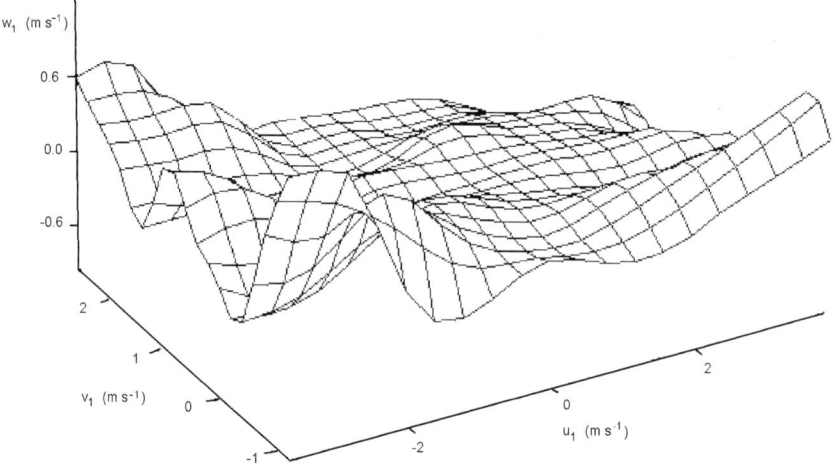

Figure 3.4. Contour plot of instrument velocity components showing the interference on air motion by the instrument tower at the Wind River site, Washington, U. S. A. (Paw U et al. 2004).

complished by firmware for the user. The geometry of the anemometer has some bearing in the way correction is made for the cross-wind effect on the sonic signal (Kaimal et al. 1990, Liu et al. 2001) and for flux loss due to pathlength averaging (Chapter 4)[3].

The base vectors of the instrument coordinate system are fixed once the position of the anemometer is known relative to some geographic reference. For example, the $\vec{e}_1$ vector may be pointing to the north, the $\vec{e}_3$ vector to the west, and the $\vec{e}_2$ vector aligned with, and in the opposite direction of, the gravitational force if the anemometer is leveled. In this sense, the instrument coordinate is an absolute one that is independent of the flow field. Micrometeorologists without exception should always archive the data of velocity statistics and flux cross products expressed in this coordinate. Although the flux cross products must undergo coordinate rotation, the velocity data themselves can be useful in many other ways. For example, the instrument velocity components can be used to determine wind direction, to infer the extent of aerodynamic interference

---

[3] Strictly, correction for flux loss due to pathlength averaging should be made with the velocity spectra in the non-orthogonal coordinate aligned with the separation direction of the transducers, not in any other coordinate (e. g., the natural wind system) unrelated to the geometry of the sonic anemometer design.

by the measurement platform (Figures 3.4 and 3.7), and to determine the orientation of the base vectors, in the instrument coordinate system, of the planar fit coordinate system (Appendix B).

3.3 Natural wind coordinate

Tanner and Thurtell (1969) define the natural wind coordinate system as a right-handed system in which the x-axis is parallel to the (30-min) mean flow with x increasing in the direction of the flow, the z-axis is normal to and pointed away from the underlying surface. It assumes that there is no correlation between the lateral and vertical velocities ($\overline{v'w'} = 0$). Transformation to this coordinate is accomplished by a two-step rotation procedure involving three rotation angles. For the reader's convenience, a brief account of their procedure is given in Appendix A. The complete description can be found in their original report and in McMillen (1988).

An obvious advantage of the natural wind coordinate is that by forcing the mean lateral and cross wind components to zero, it aligns the x axis to the streamline at the measurement point. In an idealized homogeneous flow, this serves the function of leveling the anemometer to the surface. It offers a consistent frame through time for periods when the anemometer position has been moved frequently.

If multiple sensors are deployed in the streamwise direction, by aligning the x axis with the local wind vector at each sensor location, measurements can be expressed in a common streamline coordinate. In Section 2, we suggest that in gentle terrain the streamline coordinate is the best frame to assess mass balance (Equation 3.8). Obviously, this is a formal analysis and needs to be verified by experimental tests.

Another important feature of the natural wind coordinate is that it allows online computation of the fluxes. While scalar fluxes are not particularly sensitive to tilt errors, velocity cross products in the instrument coordinate usually do not make much sense until a coordinate rotation is made. The ability to transform in real-time the velocity cross products to the streamwise momentum flux in a coordinate aligned, albeit approximately, to the surface will help the investigator detect instrument malfunction. For example, a positive covariance $\overline{u'w'}$ after rotation usually indicates problems with the sonic anemometer.

At the time of its publication, the natural wind system was intended for a surface layer in which the flow is one dimensional, and there is no wind directional shear causing cross-wind momentum flux (that is, $\overline{v'w'} = 0$). It is a suitable system for experiments conducted at ideal sites, over selected "golden days", and in homogeneous flow, fair weather conditions. In short field campaigns at a sloped site, McMillen (1988) found that rotation to the natural wind system significantly improved his results. However, the drawbacks of the system have become apparent now that the scope of micrometeorological research has been extended considerably to include non-flat sites as well as year-round, continuous monitoring. Some of the limitations can be summarized as follows:

- *Over-rotation*: By forcing the mean vertical velocity to zero for every observational period, we run the risk of over-rotation. Section 5 gives a list of examples on when this may actually occur. Over-rotation may result in a systematic bias error in the time-integrated flux.

- *Loss of information*: Most field campaigns deploy only one eddy covariance system. The theoretical advantage of aligning the coordinate with the local wind vector is no longer compelling, since it is not possible to close the mass balance with one single sensor, and is outweighed by the disadvantage of information loss. For example, a nonzero $\overline{w}$ may exist due to thermal circulation and free convection. In rolling terrain and in direction shear (in the vertical sense) flow conditions, it is not valid to assume $\overline{v'w'} = 0$. While these quantities themselves do not permit a full mass balance closure, they offer useful information on the 3-dimensional nature of the flow influencing the measurement.

- *Degradation of data quality*: It is shown that the data quality is lower for rotation into the natural wind coordinate in comparison to the planar fit method (Chapter 9). One reason for this has to do with unrealistically large rotation angles in low wind conditions. When this occurs, the z axis is no longer in a direction along which the divergence of the eddy flux is maximized. That $\overline{v'w'} \neq 0$ in advective flow also contributes to the problem. Finnigan (2004) points out that the third rotation angle (angle β, Appendix A) constrained by forcing $\overline{v'w'}$ to zero has a closure problem and often gives physically unrealistic results.

3.4   Planar fit coordinate

This is a right-handed orthogonal coordinate in which the z-axis is perpendicular to the mean streamline plane and the y-axis is perpendicular the plane in which the short-term (30 min) velocity vector $\vec{u}$ and the z axis lie. The mean streamline plane is determined from an ensemble of observations made over weeks or longer. In this system the z coordinate is fixed over the chosen period, and x and y axes are variable

Coordinate Systems and Flux Bias Error

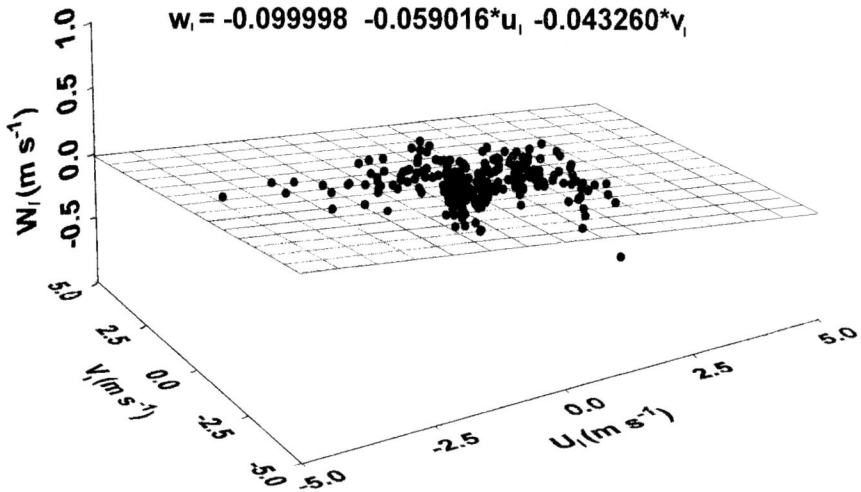

Figure 3.5. An example of planar fit regression with wind data over a maize canopy in Davis, California (Paw U et al. 2000).

with time. Strictly, the system is not a streamline coordinate because its base vectors are not aligned with the short-term mean streamline.

The steps involved in the rotation from the instrument coordinate to the planar fit coordinate are:

- Determine a period (weeks or longer) during which there was no change in the anemometer's position relative to the surface.

- Perform linear regression, $\overline{w}_1 = b_0 + b_1\overline{u}_1 + b_2\overline{v}_1$, using data from the chosen period to define a "tilted plane", or the mean streamline plane (Figure 3.5), where b_0, b_1 and b_2 are regression coefficients, and $\{\overline{u}_1, \overline{v}_1, \overline{w}_1\}$ are components of the (30-min) mean velocity in the instrument coordinate system.

- Use the regression coefficient b_1 and b_2 to determine the pitch, roll and yaw angles for rotation as in Wilczak et al. (2001) or alternatively, the base vector set that defines the three coordinate axes (Appendix B).

- Project the velocity and flux cross products into the new coordinate system.

The planar fit method overcomes some of the deficiencies of the natural wind coordinate system. The coordinate axes are not prone to the effect of instrument offset because the offset is eliminated by the least squares procedure. The z coordinate is independent of wind direction, minimizing the problem of over-rotation in the presence of the aerodynamic shadow produced by the sensor structure (Figure 3.7). By relying on a large ensemble of observations, the coordinate frame is stable through time and the x-y plane is more or less parallel to the local surface[4]. Most importantly, with the planar fit or other similar long-term coordinates, it is possible to recover information on the 2- and 3-dimensional nature of the flow field, such as the mean vertical velocity, from observations made at a single point.

In recent years the residual mean vertical velocity in the long-term coordinate has received considerable attention. Wilczak et al. (2001) consider the residual as random noise. Lee (1998), Baldocchi et al. (2000) and Paw U et al. (2000) combine the residual with the continuity equation to estimate the contribution of vertical advection to the surface layer mass balance. Finnigan (2004) views it as being indicative of low frequency contributions to the total flux. Because it is usually small in magnitude, the mean vertical velocity is very sensitive to measurement artifacts. To recover the mean vertical velocity that is truly meteorological remains a challenging task.

Several practical considerations should be kept in mind when applying the planar fit method. Every time the sonic anemometer is moved, a new base vector set or rotation angles should be determined. The rotation method assumes that the instrument offset in the vertical velocity, if any, is constant throughout the period chosen for the coordinate determination, which is made possible by the advance in the technology of sonic anemometry. Clearly, the method should not be used in situations where the offset is not stable, or when the anemometer position has been changed too frequently. In principle, the planar fit method can be implemented in the realtime computation of fluxes providing that the base vector set has been previously determined. Finally, the influences of atmospheric stability, strong winds, and change in foliage morphology on the rotation angles remain to be investigated.

4 Flux Bias Error due to Coordinate Tilt

Let us consider once again the example of one-dimensional, non-convergent wind field and horizontal homogeneity of the surface scalar source over horizontally homogeneous terrain. According to Equations 3.1-

---

[4]Sites where a systematic vertical motion exists are exceptions to this. A case in point is a forest edge where the streamline is always titled at an angle from the surface (Irvine et al. 1997, Li et al. 1990).

3.3, the eddy flux $\overline{w'c'}$ is now equivalent to the true surface-air exchange. (For simplicity, we will ignore the storage correction.) Similarly, the eddy momentum flux $\overline{u'w'}$ represents the true surface shear stress. The measurement will suffer a tilt error if it is expressed in a coordinate whose vector base $\vec{e}_2$ or the z axis is tilted from the direction normal to the surface.

4.1 Momentum flux bias

The tilt error in the momentum flux has been quantified by Wilczak et al. (2001) using the mixed-layer and surface layer similarity functions. They showed that for a 1° tilt, the error is typically greater than 10% in the surface under moderately unstable conditions and can be as large as 100% under free convection conditions. The error is probably even larger in stable conditions because of poor correlation between the streamwise and vertical velocities (Kaimal and Haugen 1969). Such a bias error is highly undesirable in the context of the Monin-Obukhov similarity because friction velocity is a velocity scale and a parameter used to define the Monin-Obukhov length and the scale for the scalar concentration. This leads to the stringent requirement of an accuracy of at least 0.1° in the internal alignment and mounting of the anemometer (Kaimal and Haugen 1969).

In the context of long-term observation of surface-air exchange of energy and materials, an accurate measurement of the momentum flux will aid gap filling and data quality control. For example, friction velocity is used to screen nighttime data for well-mixed conditions (Goulden et al. 1996). If the tilt error is large, it may be difficult to establish a friction velocity threshold for CO_2 flux. Also when applying spectral corrections to the flux, one needs an accurate measurement of stability and therefore momentum flux (Chapters 4 and 5).

4.2 Scalar flux bias

To assess the scalar flux bias error, let variables with subscript 1 denote quantities in a Cartesian coordinate tilted at an angle, α, from the correct one and variables without the subscript denote quantities in the correct coordinate. Here α is positive if the instrument is tilted into the wind and negative otherwise. The vertical eddy flux in the tilted coordinate, $\overline{w'_1 c'}$, can be expressed as

$$\overline{w'_1 c'} = \overline{w'c'} \cos(\alpha) + \overline{u'c'} \sin(\alpha). \tag{3.9}$$

Using the following approximate relationship

$$\overline{u'c'} = a\frac{\overline{u'w'}}{\overline{w'^2}}\overline{w'c'} \tag{3.10}$$

to eliminate the dependence on the horizontal eddy flux, $\overline{u'c'}$ (Lee 2004), we obtain

$$\overline{w'_1 c'} = \overline{w'c'}\cos(\alpha) + a\frac{\overline{u'w'}}{\overline{w'^2}}\overline{w'c'}\sin(\alpha). \tag{3.11}$$

Here a is an empirical constant ($a = 2.4$ and 3.3 for unstable and stable conditions, respectively). Equation 3.11 is combined with the Monin-Obukhov similarity to yield

$$\frac{\sigma_w}{u_*} = \begin{cases} 1.25(1 - 3z/L)^{1/3} & \text{for } z/L \leq 0 \\ 1.25 & \text{for } z/L > 0 \end{cases} \tag{3.12}$$

to investigate the flux bias error (Figure 3.6), where σ_w is the vertical velocity standard deviation, u_* is friction velocity, and z/L is the Monin-Obukhov stability parameter.

Figure 3.6 shows that the scalar flux is less sensitive to sensor tilt than the momentum flux, with a tilt error usually less than 5% for small tilt angles ($\alpha < 2°$). However, we should be aware of two types of systematic bias that can occur in the time integration of carbon flux. In the first, the sensor tilt angle is fixed at all times, but because the tilt error is larger in stable (nighttime) than in unstable (daytime) conditions, the overall error does not cancel out. In the second, wind direction exhibits a systematic diurnal pattern (e. g., land/sea breezes) so that the tilt angle is negative in the daytime and positive at night. This second scenario is particularly undesirable because the tilt error is of opposite sign for day versus night. If we take a typical growing season CO_2 flux of -0.5 and 0.2 $mg\,m^{-2}s^{-1}$ for daytime and nighttime, respectively, and a 4% overestimation and a 5% underestimation due to a $-2°$ and $2°$ tilt for daytime and nighttime, respectively (Figure 3.6), the bias in the monthly flux sum is estimated at 20 $g\,C\,m^{-2}$, or on the order of 10% of the annual NEE of some temperate forests.

In this simple example of 1-dimensional flow, the global property of the coordinate system is uniquely determined by the local vector basis at the measurement location. The general conclusion is applicable in weakly 2- and 3-dimensional flows. In this case, the optimal coordinate for point measurements should have its $x - z$ plane perpendicular to the local terrain surface (Figure 3.1) and the tilt error discussion should be cast in reference to this coordinate.

It should be pointed out that the above error assessment is limited to the *eddy flux* only. Errors in the overall NEE estimate caused by neglect

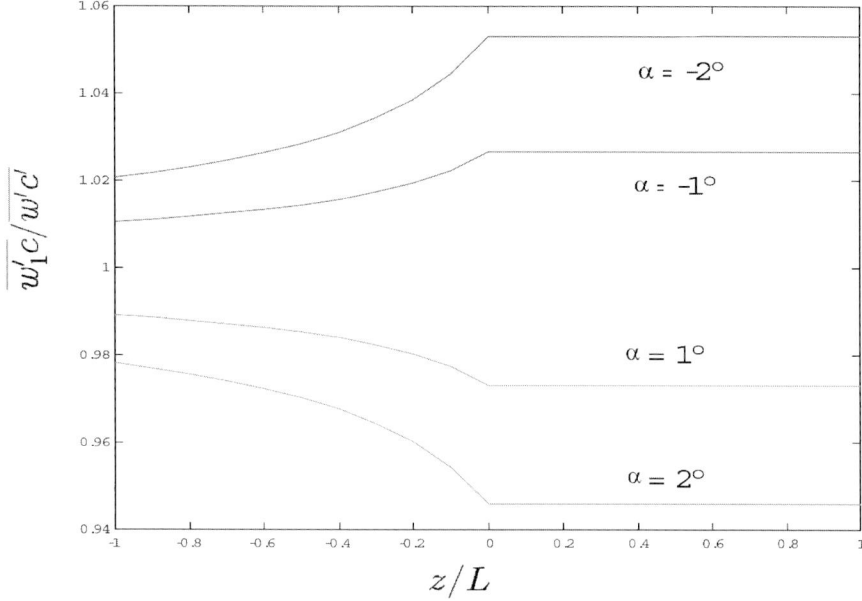

Figure 3.6. Scalar flux bias error as a function of the Monin-Obukhov stability parameter for four tilt angles.

of non-turbulent advective fluxes (e. g., $\overline{w}\,\overline{c}$) could be significantly larger in 2- and 3-dimensional flows.

5 Examples of Coordinate Tilt

Coordinate tilt can occur in several ways. The most obvious one is a physical tilt of the instrument relative to the correct coordinate frame. This can be minimized on level terrain by carefully mounting the sensor, but is unavoidable on sloped terrain because the x-y plane of an instrument leveled with respect to the geopotential is not parallel to the local terrain slope and thus is tilted from the most appropriate coordinate frame. Post-field rotation schemes attempt to remove the tilt by using wind statistics, each making a different assumption regarding the flow dynamics in the surface layer. The natural wind system assumes horizontal flow homogeneity for every observational period and thus the velocity vector is assumed to be always parallel to the surface. A coordinate system derived from an ensemble of observations assumes that the ensemble mean velocity vector is parallel to the surface, while the velocity vector over individual observational periods can intercept the surface thus allowing non-zero mean vertical velocity.

Coordinate tilt can also occur if the instrument vertical velocity has an electronic offset. [Wilczak et al. (2001) point out that offset in the instrument horizontal velocity components is not a concern.] The instrument may be perfectly aligned with the optimal coordinate frame, but in post-field rotation, such as that of Tanner and Thurtell (1969), that forces the 30-min mean vertical velocity to zero for every observation, we end up with flux and wind statistics in an incorrect reference frame. If a typical mean velocity is 2 $m\,s^{-1}$, a 5 $cm\,s^{-1}$ offset in the instrument vertical velocity is equivalent to a 1.5° tilt. This "over-rotation" will introduce a bias to the integrated carbon flux especially if wind direction changes systematically from day to night. The instrument zero offset can be measured in the field by putting the anemometer in a zero wind, anechoic chamber and be removed from the signal before coordinate rotation is performed. Care should be exercised to ensure that the zero wind chamber is not subject to differential heating as to create convective motion inside. Alternatively, the offset can be removed by a least squares regression on the assumption that it remains constant over the entire experimental period (Paw U et al. 2000; Appendix B).

Another cause of coordinate tilt arises from 2- or 3-dimensional air motion. If there is horizontal flow convergence/divergence, the (30-min) mean velocity vector will no longer parallel to the terrain surface. Once again, a tilt error will result from the mean vertical velocity being forced to zero by post-field rotation. In this regard, a coordinate system based on velocity data obtained over long periods is more robust, particularly at times of low wind speed when the natural wind system often gives unrealistically large rotation angles.

The anemometer supporting frame and the instrument tower can deflect the flow to the extent that can lead to a tilted coordinate in post-field data analysis. Figure 3.7 shows an example of this problem. The tilt factor b was determined by linear regression of the instrument mean vertical velocity, $\overline{w}_1$, against the instrument horizontal velocity, $\overline{u}_1$, as in

$$\overline{w}_1 = a + b\overline{u}_1, \qquad (3.13)$$

over successive 15° wind direction bins (Lee 1998). The sinusoidal behavior of b as a function of wind direction, expected for the ideal case of flow free of aerodynamic interference, was not observed, suggesting the aerodynamic shadow effect on the measurement. In fact, the 120°

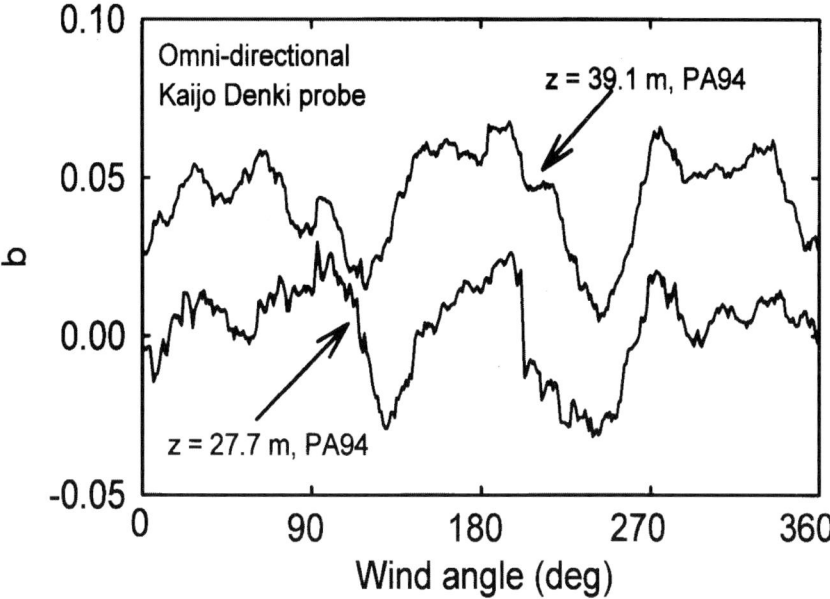

Figure 3.7. Vertical tilt factor as a function of wind direction for omnidirectional Kaijo Denki sonic anemometers at two measurements heights in a boreal aspen forest in Prince Albert, Saskatchewan, Canada.

repetitive pattern shown in Fig 3.7 corresponds to the three vertical supporting frames of the anemometer arranged 120° apart. According to Figure 3.7, by forcing the mean velocity to zero, the natural wind system can tilt the coordinate by as much as 3°. A reasonable solution to this problem is the planar fit method discussed above, which uses the data from all wind directions to determine a more stable reference frame independent of wind direction.

Finally, forcing the cross-wind momentum flux $\overline{v'w'}$ to zero may result unrealistically large rotation angles (Section 6). The tilt error in this case arises from contamination of the vertical flux $\overline{w'c'}$ by the cross-wind flux $\overline{v'c'}$ (Equation 3.17), and is usually much smaller than that arising from contamination by the streamwise flux $\overline{u'c'}$ (Equation 3.9).

6 Analysis of a Sample Dataset

6.1 Dataset

In this Section, we use a dataset obtained over the Great Mountain Forest in rolling terrain to investigate the effect of coordinate rotation

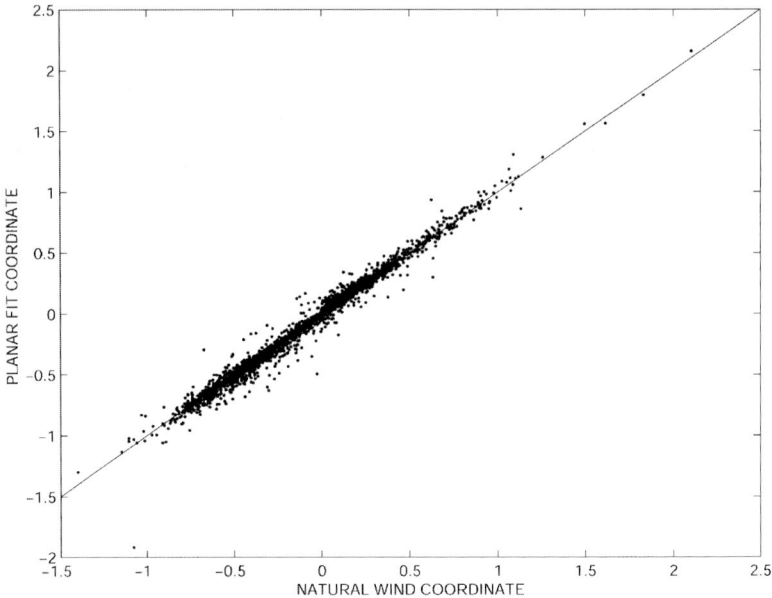

Figure 3.8. Comparison of CO_2 flux ($\text{mg m}^{-2}\,\text{s}^{-1}$) in the natural wind and planar fit coordinates. Solid line represents 1:1.

on the flux measurement. A detailed description of the site and measurement system is given by Lee and Hu (2002). Briefly, the eddy covariance system was mounted at a height of 30.4 m, roughly 10 m above the treetops. The data obtained over June to July, 1999 was used in this analysis. The 30-min velocity and flux cross product matrix was first computed in the instrument coordinate and then transformed to the natural wind coordinate system. Rotation into the planar fit coordinate was carried out in the post field analysis. Over this period, the unit vector in the direction of the z axis of the planar fit coordinate was $\{0.060, -0.078, 0.990\}$ (Appendix B). Density corrections were applied to carbon and water vapor fluxes in all three coordinate systems.

6.2 Results

Figures 3.8 and 3.9 compare the CO_2 flux and the streamwise momentum flux in the natural wind and planar fit coordinate systems. Although statistically the slope of the regression is not different from the 1:1 line, some scatter is evident. The time integrated C flux over the two month period was -84.4, -84.8 and -88.1 g C m^{-2} in the instrument, natural

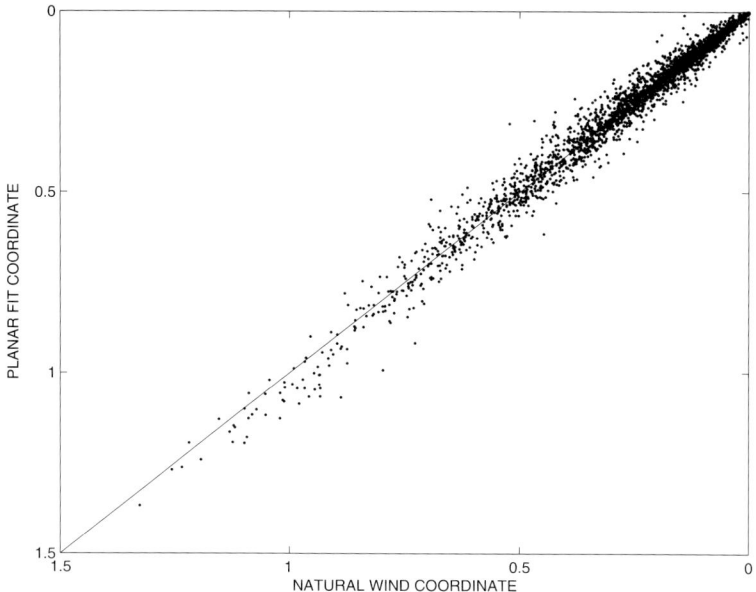

Figure 3.9. Comparison of the streamwise momentum flux ($\overline{u'w'}$, $\text{m}^2\,\text{s}^{-2}$) in the natural wind and planar fit coordinates. Solid line represents 1:1.

wind, and planar fit coordinate, respectively, with a relative difference of 4%.

Figure 3.10 shows that the cross-wind momentum flux $\overline{v'w'}$ in the planar fit coordinate is usually not negligible. This is not a surprise for a surface layer over rolling topography. That $\overline{v'w'}$ is dependent upon wind direction also suggests some tower interference with the measurement. Forcing $\overline{v'w'}$ to zero would require an additional rotation of the $z-y$ plane around the x axis by as much as 20°.

To simulate the natural wind rotation scheme, we perform one additional rotation of the velocity and flux cross products in the planar fit coordinate by forcing the cross-wind momentum flux to zero. The results are given in Figures 3.11 and 3.12. Much of the scatter in Figures 3.8 and 3.9 is eliminated by the additional rotation. The R^2 value is improved from 0.983 to 0.997 for CO_2 flux and from 0.979 to 0.991 for momentum flux. Thus the primary difference between the two coordinates is the third rotation of the natural wind system that forces $\overline{v'w'}$ to zero. A rotation angle as large as 20° is clearly not physical. Fortunately, the error caused by this rotation (rotation of the $z-y$ plane around the x axis) is much smaller than the error caused by sensor tilt

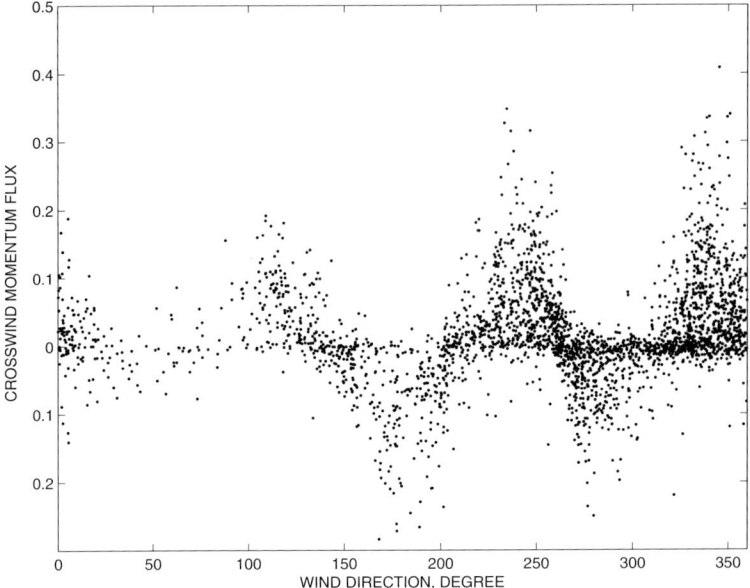

Figure 3.10. Cross-wind momentum flux ($\overline{v'w'}$, m^2 s^{-2}) in the planar fit coordinate as a function of wind direction.

in the streamwise direction discussed in Section 4. This is because the cross-wind scalar flux, $\overline{v'c'}$, and momentum flux, $\overline{v'w'}$, are much smaller than their streamwise counterparts, $\overline{u'c'}$ and $\overline{u'w'}$.

7 Conclusions

To convert measurement of wind speed, eddy flux and scalar concentration into estimates of the true surface-air exchange, we implicitly or explicitly assimilate the measurement into mathematical statements of the mass balance over a representative patch of the surface. The form of these statements depends on the coordinate system in which it is written and the coordinate system should be chosen so that the measurements can be used optimally. A comparative analysis of some candidate coordinate systems is performed, with a particular emphasis on the Cartesian and physical streamline systems.

In our theoretical analysis, we make a distinction between the vector basis, a local property of a coordinate system, and the overall coordinate frame consisting of the vector basis and coordinate lines, a global property of the flow that is determined by the flow field in three dimensions. Usually only a single tower is available as measurement platform. Such

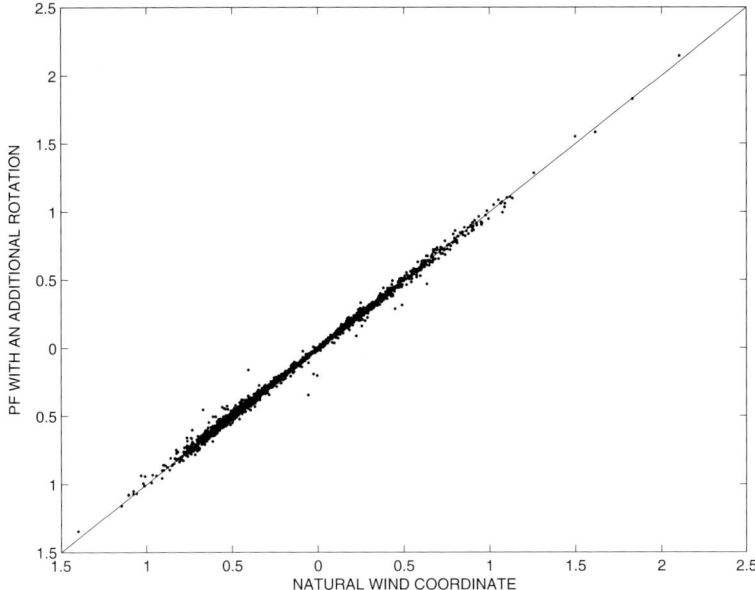

Figure 3.11. Comparison of CO_2 flux ($mg\,m^{-2}\,s^{-1}$) in the natural wind coordinate and the planar fit (PF) coordinate with one additional rotation that forces $\overline{v'w'}$ to zero. Solid line represents 1:1.

point measurements can only define the vector basis, and many components that enter into the mass balance in complex flows are severely under-sampled. A suitable coordinate frame for point measurements must optimize our estimates of the surface-air exchange using the terms we can measure, and maximize information for diagnostics purposes.

We analyze the strengths and weaknesses of three operational coordinate systems for point measurements (instrument, natural wind, and planar fit). Results of the analysis of a sample dataset shows that the cumulative C flux is 4% higher in magnitude in the planar fit coordinate than in the natural wind coordinate. The difference in the eddy fluxes in the two coordinates results primarily from the third rotation performed by the natural wind system that forces $\overline{v'w'}$ to zero.

Coordinate tilt can occur in a number of ways. Besides the obvious physical tilt of the anemometer relative to the local terrain surface, coordinate tilt can easily result from post-field data analysis. Tilt error in the eddy scalar flux $\overline{w'c'}$ arises from contamination by the streamwise flux $\overline{u'c'}$ and the cross-wind flux $\overline{v'c'}$, the former of which is much larger in magnitude. That the scalar flux tilt error is usually small for small tilt

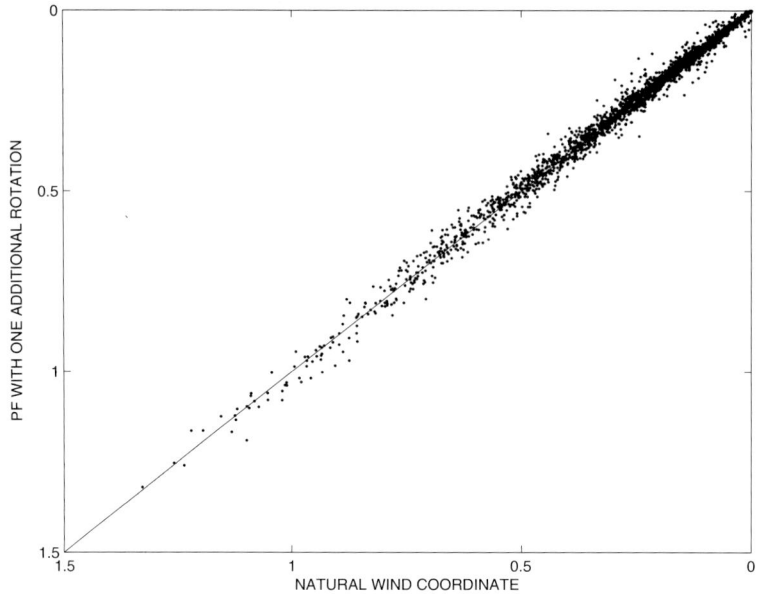

Figure 3.12. As in Figure 3.11 but for the streamwise momentum flux ($\overline{u'w'}$, m^2 s^{-2}).

angles does not negate the need for coordinate rotation. At sites where wind direction exhibits a systematic diurnal pattern, the time integrated C flux can suffer a systematic bias error on the order of 20 g C m^{-2} per month for a 2° tilt in the streamwise direction.

8 Appendix A: The Natural Wind Coordinate System

Let subscript 1 the denote velocity components and coordinate axes in the instrument coordinate. To force the mean lateral and vertical velocities to zero, we rotate through an angle η around the z_1-axis and an angle θ around the y_1-axis. The instant velocity components after the rotation, denoted with subscript 2, are

$$u_2 = u_1(\text{CT})(\text{CE}) + v_1(\text{CT})(\text{SE}) + w_1(\text{ST})$$
$$v_2 = v_1(\text{CE}) - u_1(\text{SE}) \qquad (3.14)$$
$$w_2 = w_1(\text{CT}) - u_1(\text{ST})(\text{CE}) - v_1(\text{ST})(\text{SE})$$

where

$$(\text{CE}) = \cos\eta \equiv \overline{u}_1/(\overline{u}_1^2 + \overline{v}_1^2)^{1/2}$$

$$(\text{SE}) = \sin\eta \equiv \overline{v}_1/(\overline{u}_1^2 + \overline{v}_1^2)^{1/2}$$
$$(\text{CT}) = \cos\theta \equiv (\overline{u}_1^2 + \overline{v}_1^2)^{1/2}/(\overline{u}_1^2 + \overline{v}_1^2 + \overline{w}_1^2)^{1/2}$$
$$(\text{ST}) = \sin\theta \equiv \overline{w}_1/(\overline{u}_1^2 + \overline{v}_1^2 + \overline{w}_1^2)^{1/2}$$
(3.15)

To force $\overline{w'v'}$ to zero, we must rotate the intermediate z_2-y_2 plane through an angle β. After this third rotation, we obtain

$$u = u_2$$
$$v = v_2(\text{CB}) + w_2(\text{SB})$$
$$w = w_2(\text{CB}) - v_2(\text{SB})$$
(3.16)

where

$$\text{CB} = \cos\beta$$
$$\text{SB} = \sin\beta$$

and

$$\beta = \frac{1}{2}\tan^{-1}\left[\frac{2\overline{v'_2 w'_2}}{(\overline{v'^2_2} - \overline{w'^2_2})}\right]$$

By performing Reynolds decomposition and averaging, we can determine the velocity cross products and the flux vector in the natural coordinate from those reported in the instrument coordinate. For example, the vertical flux of scalar c is

$$\overline{w'c'} = \overline{w'_2 c'}(\text{CB}) - \overline{v'_2 c'}(\text{SB})$$
(3.17)

where

$$\overline{w'_2 c'} = \overline{w'_1 c'}(\text{CT}) - \overline{u'_1 c'}(\text{ST})(\text{CE}) - \overline{v'_1 c'}(\text{ST})(\text{SE})$$
$$\overline{v'_2 c'} = \overline{v'_1 c'}(\text{CE}) - \overline{u'_1 c'}(\text{SE})$$

9 Appendix B: An Alternative Method for Rotation into the Planar Fit Coordinate

In Wilczak et al. (2001), rotation into the planar fit coordinate is accomplished by three successive steps according to pitch, roll and yaw angles. The sequence of rotation cannot be mixed. Here in the spirit of the base vector operation (Section 2), we outline an alternative approach, related to Paw U et al.'s (2000) 2-D planar fit regression. Our approach first determines the base vectors for the planar fit coordinate and then projects the measured vector quantities (velocity, flux) to each of the base vectors. This scheme relies on the straightforward vector operation and avoids the need for rotation angles and thus rotation sequence is irrelevant.

Let the unit vector set $\{\vec{i}, \vec{j}, \vec{k}\}$ define the desired right-handed orthogonal coordinate such that $\vec{i}$, $\vec{j}$ and $\vec{k}$ are parallel to its x, y and z axes, respectively[5]. Thus, the mean vertical velocity in this coordinate is the inner product of $\vec{k}$ and the mean velocity vector $\vec{\overline{u}}$

$$\overline{w} = \vec{k} \cdot \vec{\overline{u}}. \tag{3.18}$$

Substituting the component forms of the two vectors in the *instrument* coordinate

$$\vec{k} = \{k_1, k_2, k_3\}, \ \vec{\overline{u}} = \{\overline{u}_1, \overline{v}_1, \overline{w}_1 - b_0\},$$

into Equation 3.18 and solving for $\overline{w}_1$, we obtain

$$\overline{w}_1 = b_0 + b_1 \overline{u}_1 + b_2 \overline{v}_2 + \overline{w}/k_3. \tag{3.19}$$

The coefficients in Equation 3.19, b_0 (instrument offset in the vertical velocity), $b_1 (= -k_1/k_3)$ and $b_2 (= -k_2/k_3)$ are determined using a least squares regression procedure on the assumption that the last term represents "random noise". The components of $\vec{k}$ can be determined once b_1 and b_2 are known (see Matlab function unit_vector_k below).

Next we know that the y axis is perpendicular to $\vec{k}$ by definition of an orthogonal coordinate, and to $\vec{\overline{u}}$ so that after rotation the mean lateral velocity vanishes. Thus

$$\vec{j} = \vec{k} \times \vec{\overline{u}} / |\vec{k} \times \vec{\overline{u}}|. \tag{3.20}$$

Also by definition of a right-handed orthogonal coordinate, we have

$$\vec{i} = \vec{j} \times \vec{k}.$$

(Matlab function unit_vector_ij).

After all the three unit base vectors are known, the fluxes and velocity statistics can be projected easily onto the appropriate axes (Matlab functions scalar_flux and velocity_stat). For example, the vertical scalar flux is the inner product of the flux vector and vector $\vec{k}$

$$\overline{w'c'} = \{\overline{u'_1 c'}, \overline{v'_1 c'}, \overline{w'_1 c'}\} \cdot \vec{k}$$

---

[5]The vector set $\{\vec{i}, \vec{j}, \vec{k}\}$ is the same as $\{\vec{e}_1, \vec{e}_3, \vec{e}_2\}$ in the main text. We change the notation here for convenience of coding the routine.

```
% determines unit vector k (parallel to new z-axis)
% input
%    U1(:,1):  mean u1 in instrument coordinate
%      (:,2):  mean v1 in instrument coordinate
%      (:,3):  mean w1 in instrument coordinate
% output
%    k:  unit vector parallel to new coordinate z axis
%    b0: instrument offset in w1
%
function [k,b0]=unit_vector_k(U1)
% wilczak's routine
u=(U1(:,1))'; v=(U1(:,2))'; w=(U1(:,3))';
flen=length(u);
su=sum(u); sv=sum(v); sw=sum(w); suv=sum(u*v'); suw=sum(u*w');
svw=sum(v*w'); su2=sum(u*u'); sv2=sum(v*v');
H=[flen su sv; su su2 suv; sv suv sv2]; g=[sw suw svw]';
x=H\g; b0=x(1); b1=x(2); b2=x(3);
%
% determine unit vector k
k(3)=1/(1+b1^2+b2^2);
k(1)=-b1*k(3);
k(2)=-b2*k(3);
return;


% determines unit vectors i, j (parallel to new coordinate x and y axes)
%
% input
%    U1(1):  (30-min) mean u1 in instrument coordinate
%      (2):  v1 in instrument coordinate
%      (3):  w1 in instrument coordinate
%    k:  unit vector parallel to the new coordinate z-axis
% output
%    i, j:  unit vector parallel to new coordinate x and y axes
%
function [i,j]=unit_vector_ij(U1,k)
j=cross(k,U1); j=j/(sum(j.*j))^0.5; i=cross(j,k);
return;


% determines scalar flux in new coordinate
%
% input
%    u1c,v1c,w1c:  scalar flux in instrument coordinate
%    i, j, k:  unit vectors parallel to the new coordinate x, y and
% z-axes output
%    uc,vc,wc:  scalar flux in new coordinate
%
function [uc,vc,wc]=scalar_flux(u1c,v1c,w1c,i,j,k)
H=[u1c v1c w1c]; uc=sum(i.*H); vc=sum(j.*H); wc=sum(k.*H);
```

```
return;


% determines velocity statistics in new coordinate
%input
%    u:   3 by 3 matrix of cross product of the three velocity components
%         (u(1,1) = u1^u1, u(1,2)=u1^v1, etc.) in instrument coordinate
%    i, j, k:  unit vectors parallel to the new coordinate x, y and
% z-axes output
%    uu, vv, ww, uw, vw:  statistics in new coordinate
%
function [uu,vv,ww,uw,vw]=velocity_stat(u,i,j,k)
uu=i(1)^2*u(1,1)+i(2)^2*u(2,2)+i(3)^2*u(3,3)+...
2*(i(1)*i(2)*u(1,2)+i(1)*i(3)*u(1,3)+i(2)*i(3)*u(2,3));
vv=j(1)^2*u(1,1)+j(2)^2*u(2,2)+j(3)^2*u(3,3)+...
2*(j(1)*j(2)*u(1,2)+j(1)*j(3)*u(1,3)+j(2)*j(3)*u(2,3));
ww=k(1)^2*u(1,1)+k(2)^2*u(2,2)+k(3)^2*u(3,3)+...
2*(k(1)*k(2)*u(1,2)+k(1)*k(3)*u(1,3)+k(2)*k(3)*u(2,3));
uw=i(1)*k(1)*u(1,1)+i(2)*k(2)*u(2,2)+i(3)*k(3)*u(3,3)+...
(i(1)*k(2)+i(2)*k(1))*u(1,2)+(i(1)*k(3)+i(3)*k(1))*u(1,3)+...
(i(2)*k(3)+i(3)*k(2))*u(2,3);
vw=j(1)*k(1)*u(1,1)+j(2)*k(2)*u(2,2)+j(3)*k(3)*u(3,3)+...
(j(1)*k(2)+j(2)*k(1))*u(1,2)+(j(1)*k(3)+j(3)*k(1))*u(1,3)+...
(j(2)*k(3)+j(3)*k(2))*u(2,3);
return;
```

10 Acknowledgment

The first and third authors acknowledge support by the Biological and Environmental Research Program (BER), U. S. Department of Energy, through the National Institute for Global Environmental Change (NIGEC) under Cooperative Agreement No. DE-FC03-90ER61010. Additional support was provided by the U. S. National Science Foundation through grant ATM-0072864 (to the first author).

11 References

Batchelor, G. K.: 1967, *An Introduction to Fluid Mechanics*, Cambridge University Press, New York.

Baldocchi, D., Finnigan, J., Wilson, K., Paw U, K. T.: 2000, 'On measuring net ecosystem carbon exchange over tall vegetation in complex terrain', *Bound.-Layer Meteorol.* **96**, 257-291.

Bradshaw, P.: 1973, 'Effects of streamline curvature on turbulent flow', *AGARDograph No. 169*, National Technical Information Service, US Dept. of Commerce, pp. 125.

Ferziger, J. H., Peric, M.: 1997, *Computational Methods for Fluid Dynamics*, Springer-Verlag, Berlin.

Finnigan, J. J.: 2004, 'A re-evaluation of long-term flux measurement techniques. Part II: coordinate systems', *Bound.-Layer Meteorol.* in review.

Finnigan, J. J., Clements, R., Malhi, Y., Leuning, R., Cleugh, H.: 2003, 'A reevaluation of long-term flux measurement techniques. Part I: averaging and coordinate rotation', *Bound.-Layer Meteorol.* **107**, 1-48.

Finnigan, J. J.: 1990, 'Streamline coordinates, moving frames, chaos and integrability in fluid flow', *Topological Fluid Mechanics, Proc. IUTAM Symp. Topological Fluid Mechanics*, Eds Moffat, H. K., Tsinober A., Cambridge University Press, Cambridge, 64-74.

Finnigan, J. J.: 1983, 'A streamline coordinate system for distorted two-dimensional shear flows', *J. Fluid Mech.* **130**, 241-258.

Finnigan, J. J., Bradley, E. F.: 1983, 'The turbulent kinetic energy budget behind a porous barrier: an analysis in streamline coordinates', *J. Wind Eng. Ind. Aerodyn.* **15**, 157-168.

Geissbuhler, P., Siegwolf, R., Eugster, W.: 2000, 'Eddy covariance measurements on mountain slopes: the advantage of surface-normal sensor orientation over a vertical set-up', *Bound.-Layer Meteorol.* **96**, 371-392.

Goulden, M. L., Munger, J. W., Fan, S.-M., Daube, B. C., Wofsy, S. C.: 1996, 'Measurements of carbon sequestration by long-term eddy covariance methods and a critical evaluation of accuracy', *Global Change Biology* **2**, 169-183.

Howarth, L.: 1951, 'The boundary-layer in three dimensional flow. Part I: derivation of the equations for flow along a general curved surface', *Phil. Mag.* **42**, 239-243.

Irvine, M. R., Gardiner, B. A., Hill, M. K.: 1997, 'The evolution of turbulence across a forest edge', *Bound.-Layer Meteorol.* **94**, 467-497.

Kaimal, J. C., Finnigan, J. J.: 1994, *Atmospheric Boundary Layer Flows: Their Structure and Measurement*, Oxford University Press, New York.

Kaimal, J. C., Gaynor, J. E., Zimmerman, H. A., Zimmerman, G. A.: 1990, 'Minimizing flow distortion errors in a sonic anemometer', *Bound.-Layer Meteorol.* **53**, 103-115.

Kaimal, J. C., Haugen, D. A.: 1969, 'Some errors in the measurement of Reynolds stress', *J. Appl. Meteorol.* **8**, 460-462.

Lee, X.: 2004, 'Forest-atmosphere exchanges in non-ideal conditions: the role of horizontal eddy flux and its divergence' *Forest at the Land-Atmosphere Interface* (Mencuccini, M. et al. Eds), CAB International, pp145-157.

Lee, X., Hu, X.: 2002, 'Forest-air fluxes of carbon and energy over non-flat terrain', *Bound.-Layer Meteorol.* **103**, 277-301.

Lee, X.: 1998, 'On micrometeorological observations of surface-air exchange over tall vegetation', *Agric. Forest Meteorol.* **91**, 39-49.

Li, Z., Lin, J. D., Miller, D. R.: 1990, 'Air flow over and through a forest edge: a steady state numerical simulation', *Bound.-Layer Meteorol.* **51**, 179-197.

Liu, H. P, Peters, G., Foken, T.: 2001, 'New equations for sonic temperature variance and buoyancy heat flux with an omnidirectional sonic anemometer', *Bound.-Layer Meteorol.* **100**, 459-468.

Massman, W. J., Lee, X.: 2002, 'Eddy covariance flux corrections and uncertainties in long-term studies of carbon and energy exchanges', *Agric. Forest Meteorol.* **113**, 121-144.

McMillen, R. T.: 1988, 'An eddy correlation technique with extended applicability to non-simple terrain', *Bound.-Layer Meteorol.* **43**, 231-245.

Paw U, K. T., Baldocchi, D., Meyers, T. P., Wilson, K. B.: 2000, 'Correction of eddy-covariance measurements incorporating both advective effects and density fluxes', *Bound.-Layer Meteorol.* **97**, 487-511.

Paw U, K. T., Falk, M., Suchanic, T. H., Ustin, S. L., Chen, J., Park, Y.-S., Winner, W. E., Thomas, S. C., Hsiao, T. C., Shaw, R. H., King, T. S., Pyles, R. D., Schroeder, M., Matista, A. A.: 2004, 'Carbon dioxide exchange between an old growth forest and the atmosphere', *Ecosystems*, in press.

Pielke, R. A.: 1984, *Mesoscale Meteorological Modeling*, Academic Press, New York.

Raupach, M. R.: 2001 'Inferring Biogeochemical sources and sinks from atmospheric concentrations: general considerations and applications in vegetation canopies', In Shulze, E-D., Heimann, M., Harrison, S., Holland, E., Lloyd, J., Prentice, I. C., Schimel, D. (Eds) *Global Biogeochemical Cycles in the Climate System* , Academic Press, 41-59.

Sakai, R. K., Fitzjarrald, D. R., Moore, K. E.: 2001, 'Importance of low-frequency contributions to eddy fluxes observed over rough surfaces', *J. Appl. Meteorol.* **40**, 2178-2192.

Tanner, C. B., Thurtell, G. W.: 1969, 'Anemoclinometer measurements of Reynolds stress and heat transport in the atmospheric surface layer', *Research and Development Tech. Report ECOM 66-G22-F to the US Army Electronics Command*, Dept. Soil Science, Univ. of Wisconsin, Madison, WI.

Wilczak, J. M., Oncley , S. P., Sage, S. A.: 2001, 'Sonic anemometer tilt correction algorithms', *Bound.-Layer Meteorol.* **99**, 127-150.

Zeman, O., Jensen, N. O.: 1987, 'Modification of turbulence characteristics in flow over hills", *Quart. J. Roy Meteorol. Soc.* **113**, 55-80.

Chapter 4

UNCERTAINTY IN EDDY COVARIANCE FLUX ESTIMATES RESULTING FROM SPECTRAL ATTENUATION

William Massman, Robert Clement
wmassman@fs.fed.us

Abstract

Surface exchange fluxes measured by eddy covariance tend to be underestimated as a result of limitations in sensor design, signal processing methods, and finite flux-averaging periods. But, careful system design, modern instrumentation, and appropriate data processing algorithms can minimize these losses, which, if not too large, can be estimated and corrected using any of several different approaches. No flux-correction method is perfect, however, so methodological uncertainties are inevitable. This study addresses the uncertainties in surface flux measurements that have been corrected for spectral attenuation with the transfer function approach. The sources of the errors in the estimates of flux attenuation examined here include the (flux-averaging) period-to-period variablity of cospectra, the departure of real cospectra from presumed smooth curves, the inherent variability in maximum frequency (f_x) of the frequency weighted cospectra, and possible imprecision in instrument related time constants. A method is proposed to estimate the uncertainty resulting from combining these effects. Also included in this study are a general mathematical relationship to describe spectra or cospectra, comparisons of observed cospectra for cospectral similarity, and discussions about including high-pass filters (associated with the flux-averaging procedures) when accounting for low frequency losses.

1 General Issues Regarding Flux Attenuation

Even the most carefully designed and deployed eddy covariance system will not be able to completely sample all flux-carrying turbulent atmospheric eddies. As a result, all eddy covariance systems tend to un-

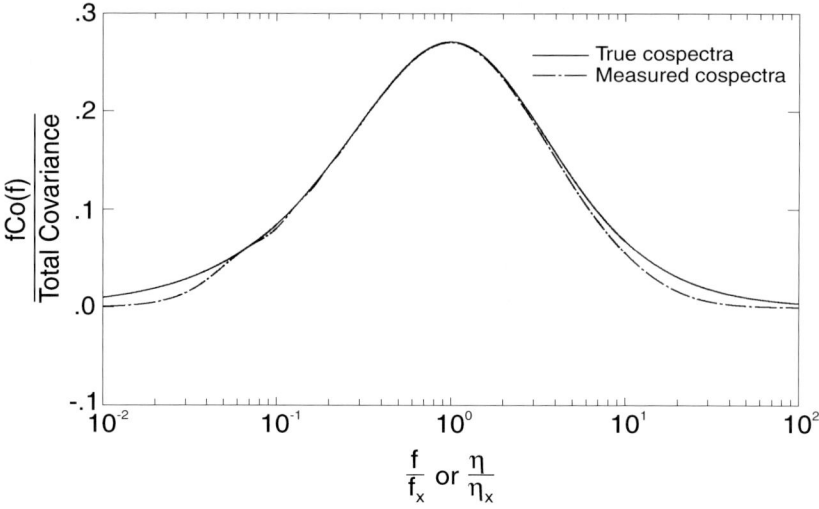

Figure 4.1. A comparison of a hypothetical frequency-weighted atmospheric cospectrum (solid line) with one measured by an eddy covariance system (dash-dot). The cospectral attenuation at high frequencies is a result of sensor line averaging, separation, and response times. The low frequency loss is a consequence of the block averaging associated with calculating time averaged fluxes. The total flux loss is related to the difference in the areas under each of the cospectral curves. Here f is frequency, f_x is the frequency at which $fCo(f)$ reaches its maximum value, and η and η_x are their respective nondimensionalized forms. Further discussions can be found in sections 2.1 and 3.2.

derestimate the true atmospheric fluxes. This bias results from the physical limitations in the size, separation distances, and response times of the sensors, the electronic filters used to reduce noise associated with an instrument's output signal, and the data processing algorithms intended to separate the turbulent fluctuations and their means. In general terms, the physical limitations of the instruments and any electronic filters tend to limit a system's ability to resolve the smallest eddies. Whereas, the mean-removal and flux-averaging methods restrict a system's ability to sample the largest eddies, a consequence of finite flux-averaging periods. Consequently, all eddy covariance systems are bandwidth limited and all measured fluxes are attenuated at both high and low frequencies (see Figure 4.1). No eddy covariance instrument or system is free of these shortcomings, but by careful design and the use of modern instrumentation and electronics many of these effects can be reduced and, within reason, quantified and corrected. To date there have been several approaches developed to deal with sensor related attenuation effects. In most cases these approaches are variants of two broad methods: the transfer function approach and in-situ methods.

Early attempts to deal with instrument-related attenuation largely focused on developing transfer functions to describe how sonic and scalar sensor designs (sensor separation lengths and geometrical shapes) might affect the high frequency portion of spectra and cospectra measured with these instruments (e. g., Gurvich 1962, Kaimal et al. 1968, Silverman 1968, Wyngaard 1971, Hicks 1972, Horst 1973). In general, transfer functions, such as those just cited, have been developed from knowledge of atmospheric turbulence and the technology and physical principles underlying the operation of the instrumentation. But, they have also been empirically determined, as Laubach and Teichmann (1996) did for water vapor tube sorption effects. By comprehensively applying spectral transfer functions to a collection of fast response instruments Moore (1986) initiated the transfer function method for estimating spectral loss associated with eddy covariance systems. More recently Horst (1997 2000) and Massman (2000, 2001) have extended Moore's (1986) approach by expanding it to include transfer functions appropriate to newer instrumentation (e. g., Massman 1991) and by including low frequency attenuation associated with flux sampling and averaging procedures (e. g., Kaimal et al. 1989, Rannik and Vesala 1999, Rannik 2001). The transfer function method, useful as a aid in minimizing and recovering flux loss related to the design of eddy covariance systems and their data handling and flux-processing algorithms, can be summarized by the following equation:

$$\frac{\overline{(w'\beta')}_m}{\overline{w'\beta'}} = \frac{\int_0^\infty \left[1 - \frac{\sin^2(\pi f T_b)}{(\pi f T_b)^2}\right] H(f) Co_{w\beta}(f) df}{\int_0^\infty Co_{w\beta}(f) df} \quad (4.1)$$

where w' and β' are the fluctuations of vertical velocity and either the horizontal wind speed or scalar concentration; $\overline{(w'\beta')}_m$ is the measured covariance and $\overline{(w'\beta')}$ is the true or unattenuated flux; f is frequency; $H(f) = \prod_{i=1}^N H_i(f)$ is the product of all the appropriate transfer functions associated with high frequency attenuation; $Co_{w\beta}(f)$ is the one-sided cospectrum; T_b is the block averaging period; and $[1 - \sin^2(\pi f T_b)/(\pi f T_b)^2]$ is the transfer function associated with block-averaging. This last transfer function or filter accounts for the low frequency spectral attenuation (Figure 4.1) and, as mentioned previously, is a consequence of the need to use a finite flux averaging period and the specific method used to separate the turbulent fluctuations from their mean.

The advantages of the transfer function method are that it is fairly comprehensive, it is independent of any measured fluxes, and it can be reasonably accurate (Laubach and McNaughton 1999). All that is required to use this method are the flux averaging period, the transfer functions, and a model of the cospectra. Unfortunately, the need for a cospectral model is also a weakness of the transfer function method. Usually the cospectra are modeled as relatively smooth (continuous) functions. For example, Moore (1986) used Kaimal's et al. (1972) smooth flat terrain spectra and cospectra and Horst (2000) used a much simpler formulation for the cospectra that provided a very good approximation to the flat terrain cospectra. But, smooth spectra and cospectra are not typical of the atmospheric surface layer. In fact, virtually all observed half hourly cospectra display significant variablity and virtually none of them resemble a smooth shape (e. g., Laubach and McNaughton 1999).

Another shortcoming with the transfer function method is that it is mathematically complicated and therefore numerically intensive. Although modern PCs have alleviated this problem somewhat, still some of the transfer functions are difficult to evaluate in their original formulations and the numerical techniques required to do the integration are not necessarily obvious. Moore (1986) suggested approximating the transfer functions by simpler functions, but retained the computational aspects of Equation 4.1. Horst (1997) and Massman (2000), on the other hand, have suggested a much simpler analytical approach as an alternative to the direct use of Equation 4.1. For relatively small (high frequency) attenuation effects the analytical model is an extremely good substitute for Equation 4.1 (Massman 2000, 2001). But for relatively larger amounts of (high frequency) attenuation, Massman (2000, 2001) indicates that the analytical method can significantly underestimate Equation 4.1.

A final concern with the transfer function method is that the associated correction factors can become quite large (e. g., Villalobos 1996). This is particularly true at night or during low wind speeds (Massman 2001). Although it is less clear how significant a large correction factor, one greater than about 1.5 for example, may have on the long-term carbon budget because they are typically associated with small fluxes (Massman and Lee 2002).

Unlike the transfer function approach, in situ methods do not require smooth models of atmospheric cospectra. Fundamental to these methods is the assumption of cospectral similarity between scalar fluxes. Application of this method, in the most general terms, requires taking the ratio of a reference flux (usually assumed to have no attenuation) to an attenuated reference flux. This ratio is then used as a correction factor for a measured and cospectrally similar scalar flux. This basic approach has several variants. The pass band covariance approach, first proposed by Hicks and McMillen (1988), basically assumes cospectral similarity in the central portion of any measured scalar cospectra and that this central cospectral band is not significantly influenced by either low or high frequency attenuation effects. This approach has been used in several different ways to reconstruct or correct measured fluxes (e. g., Mestayer et al. 1990, Verma et al. 1992, Horst et al. 1997, Aubinet et al. 2001). A second variant uses the heat flux as both the reference flux and the degraded reference flux. This approach has been used by, among others, by Koprov et al. (1973), Lee and Black (1994), Kristensen et al. (1997), Laubach and McNaughton (1999), and Villalobos (2001) to study and correct for attenuation resulting from lateral sensor separation effects. However, Goulden et al. (1997), using a recursive low pass digital filter, applied the degraded heat flux approach for real time correction of CO_2 fluxes measured with a closed-path sensor.

While these in situ methods do not suffer from issues regarding modeled cospectra like the transfer function approach does, they, nevertheless, have drawbacks that introduce uncertainty into the fluxes when they are used to estimate flux corrections. First, cospectral similarity is not guaranteed (e. g., Katul and Hsieh 1999). Differing distributions for the sources and sinks of CO_2, H_2O, and heat will produce cospectral dissimilarity between these scalars. Second, this method does not include high frequency corrections to the reference flux, which is usually taken to be the heat flux as measured by sonic anemometry, or for line averaging effects on the vertical velocity signal. These will result in the underestimation of the corrected fluxes. This error, however, should be a relatively small (e. g., 2-6%, Massman 2001) that will likely vary with wind speed and atmospheric stability. Third, the reference flux must be large enough to be adequately resolved by the eddy covariance system; otherwise it cannot be used to define the correction factor. This often makes nighttime flux corrections problematic. (Of course flux measurement at night is difficult for other reasons as well.) Fourth, calibrating the in situ method (appropriate to closed path eddy covariance systems) is, by its nature, somewhat imprecise. For example, in order to digitally degrade the temperature flux so that it will emulate the attenuation associated with a closed-path CO_2 system, it is necessary to provide an appropriate time constant. How this time constant is determined is critical to this method. If the time constant is found by comparing spectra of temperature and CO_2 or some other trace gas, then attenuation effects resulting from any spatial separation between the CO_2 intake tube and the sonic anemometer will have to be treated separately. Furthermore, digitally degrading the temperature flux using this time constant will also introduce a phase between the sonic signal and the degraded temperature signal, which may be different than the phase between the sonic signal and the CO_2 signal. Depending on the nature of these two different phase shifts, this could result in either overestimating or underestimating the high frequency spectral correction factor for CO_2 flux. On the other hand, if this time constant is determined by comparing the observed scalar cospectra, then in principle the time constant can be influenced by effects associated with the lateral and longitudinal separations between the CO_2 intake tube, the temperature sensor, and the sonic anemometer. Therefore, empirically determined time constants, often used to characterize closed path eddy covariance systems, can be a source of uncertainty for spectral corrections. This is much less of an issue with open path systems for which the time constants are much better defined. Fifth, low frequency corrections are not included with in situ methods, which may or may not be a serious problem because the low frequency corrections are likely to need special treatment, as discussed again in a later section.

In summary any method used for quantifying and correcting eddy covariance fluxes for spectral loss will also carry some inherent uncertainty. Such methods must employ simplifying assumptions, which will be violated to varying degrees from one eddy covariance system or instrument to another and from one flux sampling period to another. Nevertheless without such assumptions quantifying and compensating for spectral loss in eddy covariance flux data becomes prohibitive or impossible. It is, therefore, beneficial to examine the uncertainties associated with these assumptions and, if possible, to quantify them.

Specifically, this study develops a simple expression (or method) for estimating the uncertainty in the spectral correction factors derived from the transfer function method. This expression, based on Massman's (2000, 2001) analytical model for spectral corrections, includes the major aspects of cospectral variability. We focus on the transfer function method because it lends itself to error estimation more easily than does the in situ method. This uncertainty analysis is applied to two hypothetical systems: an open path system and a closed path system with an empirically determined time constant. We also examine some aspects of cospectral similarity by introducing and using a general mathematical form for (ideal) spectra and cospectra and comparing heat and momentum cospectra at an AmeriFlux and a CarboEurope flux site. Ultimately, we hope this study will provide an extended example of how to develop and assess spectral correction factors that can be used at any site and with any eddy covariance system. The next section uses a formal error analysis to define the sources of uncertainties associated with the transfer

Flux Loss from Spectral Attenuation 73

function method for estimating spectral corrections. The third section describes how the uncertainty analysis is applied to observed data.

2 Sources of Uncertainties for the Transfer Function Method

For the sake of consistency and to emphasize the relationship with Massman's (2000, 2001) approach for spectral corrections we will, as appropriate, employ his notation throughout this study. We will also henceforth reference his papers as M21.

For spectral correction methods based on transfer functions the major source of uncertainty is the variability in the cospectra from one flux averaging period to another. This cospectral variation has three aspects: (i) variability in the frequency, f_x, at which the frequency-weighted cospectra, $fCo(f)$, reaches a maximum value, (ii) variations in how relatively broad or peaked $fCo(f)$ might be (Horst 1997; M21), and (iii) departures of any observed cospectra during an averaging period from the assumed smooth shape. A fourth, but less significant, source of uncertainty results from (iv) uncertainties in an eddy covariance system's time constants. Issues (i), (ii), and (iv) are treated within the analytic framework developed by M21. The third concern is treated numerically. The next subsection defines the concepts related to (i) and (ii) in terms of a universal mathematical expression for cospectra and spectra. After that section 2.2 develops an uncertainty analysis using M21's analytical model. Later sections discuss closed path systems and the associated uncertainties in system time constants and also describe how the analytical results are generalized to include the issues related to (iii).

2.1 A general mathematical expression for spectra and cospectra

A fairly general (smooth) mathematical expression or model of cospectra that will be used with transfer function based methods to estimate spectral correction factors is

$$fCo(f) = A_0 \frac{f/f_x}{[1 + m(f/f_x)^{2\mu}]^{\frac{1}{2\mu}(\frac{m+1}{m})}} \quad (4.2)$$

where f is frequency (Hz), A_0 is a normalization parameter (discussed in more detail below), m is the (inertial subrange) slope parameter, and μ is the broadness parameter. To describe cospectra, which are normally characterized by a -7/3 power law in the inertial subrange, m must be 3/4. To describe spectra $m = 3/2$ results in a -5/3 power law.

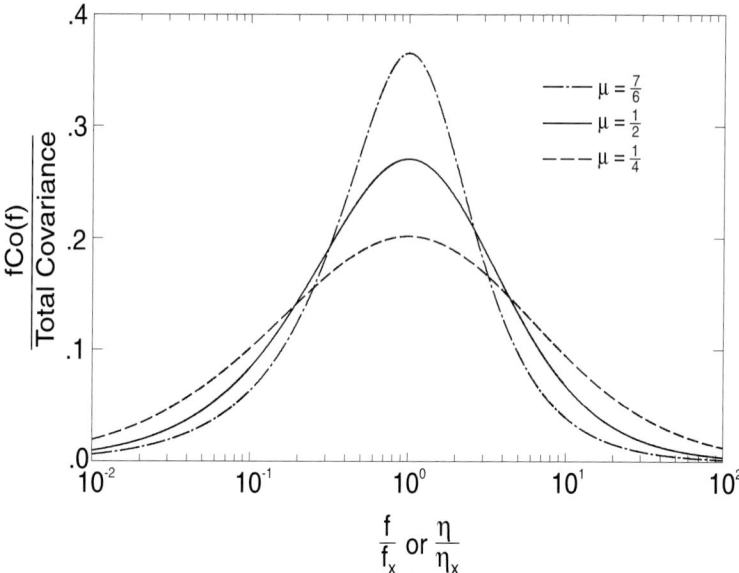

Figure 4.2. Influence of the broadness parameter μ on the shape of the cospectra modeled by Equation 4.1. For $\mu = 7/6$ Equation 4.1 approximates the flat terrain stable-atmosphere cospectra of Kaimal et al. (1972). For $\mu = 1/2$ it approximates their cospectra for an unstable atmosphere. All cospectra obey a -7/3 power law in the inertial subrange.

Equation 4.2 is a generalization of the models of spectra and cospectra discussed by Busch (1973) and Kristensen et al. (1997). It can be used to describe either normalized or unnormalized spectra or cospectra. A normalized spectra or cospectra would have unit area, i. e., $\int_0^\infty Co(f)df/$(total covariance) $= 1$. If $Co(f)$ is not divided by the total covariance the cospectra would be unnormalized. Clearly, it is possible to employ the normalization condition to eliminate A_0 from Equation 4.2. For example, the condition of unit area for a normalized spectra or cospectra yields

$$A_0 \frac{B(\frac{1}{2\mu}, \frac{1}{2\mu m})}{2\mu(m^{1/2\mu})} = 1$$

where $B(x, y)$ is the complete beta function (e. g., Spanier and Oldham 1987). However, for the present study A_0 will be taken as a free parameter. This is done largely for ease of computation, otherwise obtaining the parameters f_x, μ, and m by nonlinear regression on observed cospectra would be considerably more complicated both mathematically and numerically.

An example of the broadness parameter is shown in Figure 4.2. For $m = 3/4$ and $\mu = 7/6$ Equation 4.2 closely approximates the flat terrain stable-atmosphere frequency-weighted cospectra of Kaimal et al. (1972); whereas, for $\mu = 1/2$ it approximates their cospectral model of an unstable atmosphere. In general, Figure 4.2 shows that relatively narrow or peaked cospectra are associated with larger values of μ and that relatively broad cospectra are associated with smaller values of μ.

2.2 Analytical expression for estimating uncertainty

One of the benefits of any analytical model is that it can be used in a formal error analysis to estimate model uncertainty resulting from errors or uncertainties associated with the model's parameters. M21 shares this advantage. In its simplest form his model yields the following expression for the ratio of the measured covariance, $\overline{(w'\beta')}_m$, to the true or spectrally unattenuated covariance, $\overline{(w'\beta')}$:

$$\frac{\overline{(w'\beta')}_m}{\overline{w'\beta'}} = [\frac{b^\alpha}{(b^\alpha+1)}][\frac{b^\alpha}{(b^\alpha+p^\alpha)}][\frac{1}{p^\alpha+1}] \quad (4.3)$$

where $b = 2\pi f_x \tau_b$ and τ_b = the equivalent time constant associated with block averaging (Massman 2000); $p = 2\pi f_x \tau_e$ and τ_e = the equivalent time constant associated with all high frequency attenuation (Massman 2000); and α is the broadness parameter for M21's analytical model. [Note that α is really an adjustment factor to M21's analytical model that improves the quality of the correction factor for relatively broader cospectra. It is not the same as μ, but it is related to μ and is associated with cospectra for which $\mu \leq 0.5$.] This expression assumes that recursive filtering is not used when separating the mean and fluctuating quantities so that the 'a' term of Massman (2000) is not necessary (e. g., Massman 2001).

It is useful for the present study to simplify Equation 4.3, which can be done by noting that for most applications the flux averaging period far exceeds the time constants associated with high frequency attenuation effects, i. e., $\tau_b \gg \tau_e$. This allows the middle term on the right hand side of Equation 4.3 to be approximated by unity, i. e., $\left[\frac{b^\alpha}{(b^\alpha+p^\alpha)}\right] \approx 1$. The resulting correction factor, F, is the inverse of the simplified Equation 4.3 and is given next.

$$F = \frac{\overline{w'\beta'}}{\overline{(w'\beta')}_m} = [1 + \frac{1}{(2\pi f_x \tau_b)^\alpha}][1 + (2\pi f_x \tau_e)^\alpha] \quad (4.4)$$

This expression can be used to evaluate the expected error, ΔF, in the correction factor, F, resulting from period to period cospectral variations in f_x and α and from uncertainties in τ_e. Following Scarborough (1966) or Coleman and Steele (1999) the expected error is defined as

$$\Delta F = \sqrt{\left[\frac{\partial F}{\partial f_x}\Delta f_x\right]^2 + \left[\frac{\partial F}{\partial \alpha}\Delta \alpha\right]^2 + \left[\frac{\partial F}{\partial \tau_e}\Delta \tau_e\right]^2} \qquad (4.5)$$

where Δf_x and $\Delta \alpha$ are the uncertainties associated with f_x and α and $\Delta \tau_e$ is the uncertainty associated with τ_e. Usually τ_b is known with sufficient precision that there is no need to include any associated uncertainty. The uncertainties associated with f_x and α can be evaluated from data, as outlined in the next section. Parameter $\Delta \tau_e$ is considered only for the closed path system, which is also discussed later. For an open path system with small separation distances $\Delta \tau_e$ is considered negligible. [Note that the use of τ_e in M21's model for spectral corrections does introduce some uncertainties into the spectral estimates; however they result more from mathematical approximations than from random variations. Consequently, these uncertainties are likely to produce biases (underestimations) of the true spectral correction.] Using Equation 4.4 to evaluate the partial derivatives and then setting the parameter $\alpha = 1$ in the resulting expression yields the following expression for the relative uncertainty in the correction factor, $\Delta F/F$:

$$\frac{\Delta F}{F} = \{[(1+\frac{1}{b})p - \frac{1+p}{b}]^2[\frac{\Delta f_x}{f_x}]^2$$
$$+[(1+\frac{1}{b})p\ln(p) - \frac{1+p}{b}\ln(b)]^2[\Delta \alpha]^2$$
$$+[(1+\frac{1}{b})p]^2[\frac{\Delta \tau_e}{\tau_e}]^2\}^{0.5}/[(1+p)(1+\frac{1}{b})] \qquad (4.6)$$

The broadness parameter, α, was set to 1 in this expression because its observed range variation, as discussed later, is not very great so that explicitly including it in Equation 4.6 is not necessary and because we wish to keep this expression as simple as possible.

Equation 4.6 can be interpreted in two ways. First, as previously emphasized, this expression defines the relative uncertainty in the spectral correction factor, F, used with some eddy covariance systems. Second, and perhaps more importantly, it also represents the uncertainty in flux estimates that use transfer function based methods to spectrally correct the measured covariances. This second interpretation follows from the fact that the correction factors are applied by multiplying the measured covariance by F. Nevertheless, this second interpretation is broader and more inclusive than what might be inferred from the assumptions underpinning its derivation. As discussed below, Equation 4.6 is fairly robust and can be generalized to include other sources of uncertainty associated with spectral correction factors. The next section discusses how the model parameters f_x, Δf_x, α, $\Delta \alpha$, τ_e, and $\Delta \tau_e$ are evaluated from observed data and how Equation 4.6 is generalized and applied to the measured covariances.

3 Application of Uncertainty Analysis to Observed Data

3.1 Site description and data handling preliminaries: An AmeriFlux site

Data used in this study were obtained between January 2 and January 5, 2001 at a subalpine site within the larger Glacier Lakes Ecosystem Experiments Site (GLEES), located in the Rocky Mountains of southern Wyoming about 70 km west of Laramie, Wyoming, USA. This AmeriFlux eddy covariance site has an elevation of 3158 m and its location is [41°21′56.3″N, 106°14′22.6″W]. The mean topographical slope from about 2 km upwind from the eddy covariance tower is about +3 to +5°. Beyond that the slope becomes quite steep as the land rises to a mountain peak of about 3430 m. For the purpose of measuring fluxes, the general region should be considered micrometeorologically complex. During the time of this experiment the area was snow covered. The snow depth in the clearing surrounding the eddy covariance tower was about 1.1 m and within the nearby wooded area it was about 0.7 m.

At GLEES the sonic anemometer is $z_{ref} = 27.1$ m above the ground surface and, being aligned with the dominant wind direction, is pointed nearly due west (268°). Year round the winds almost exclusively fall within the sector between 230° and 310°. This sector is forested and covered by Engelmann spruce (*Picea engelmannii* (Parry) Englem.) and subalpine fir (*Abies lasiocarpa* [Hook.] Nutt.), which range in height (h) between 15 and 20 m and in age between 250 and 450 years old. However, the region generally is patchy and also includes seasonally wet meadows, open areas within the forest, and several lakes of various sizes. During the winter months the half-hourly mean winds can vary between about 2 m s^{-1} and 25 m s^{-1}, with an mean ensemble average of about 10 m s^{-1}. The zero-plane displacement height, d, for this site is assumed to be 13 m (or $d \approx 0.75h$). The mean temperature is about -10°C and the mean pressure is about 69 kPa. As a result of solar heating of the trees and needles the sensible heat fluxes during the daylight can be large ($\sim$300 W m^{-2}), as can the latent heat fluxes ($\sim$200 W m^{-2}), which are driven by sublimation of the snow. CO_2 fluxes during the winter are low ($\sim$0.02 mg m^{-2}s^{-1}) and are a combination of efflux from the snow surface and from the tree boles. Clearly the source distributions for heat, CO_2, and water vapor fluxes are quite different at GLEES. In spite of the relatively high heat and vapor fluxes atmospheric stability at GLEES during the winter is assumed to be neutral during the January 2001 study period. This is because GLEES is very windy and extremely turbulent almost continuously during the winter months so that $z_{ref}/L = 0$ virtually 24 hours a day. Here L denotes the Monin-Obukhov length.

The turbulence data were obtained with an ATI sonic anemometer (thermometer), model no. SATI/3VX and an open-path NOAA ATDD infrared CO_2/H_2O sensor (Auble and Meyers 1992). Also included as part of this study is a fast response pressure sensor for measuring static turbulent pressure fluctuations (Cook and Bedard 1971, Nishiyama and Bedard 1991, Wyngaard et al. 1994), which is treated in this study as another atmospheric scalar. The sonic has a path length of 0.15 m. The open-path CO_2/H_2O sensor has a path length of 0.20 m and is mounted 0.25 m below the center of the sonic. Relative to the sonic the open-path sensor was displaced laterally by 0.051 m and longitudinally (behind the sonic) by 0.22 m. The pressure probe is connected to a differential pressure sensor by a 6.1 m long garden hose with a diameter of 0.025 m. The probe is positioned 0.33 m above the center of the sonic and has a path length of 0.045 m. It is displaced laterally by 0.051 m relative to the sonic and located 0.356 m behind the sonic.

For the purposes of the present work, we do not include any possible attenuation associated with vertical displacements (e. g., Kristensen et al. 1997) or possible tube effects on the pressure fluctuations (e. g., Aydin 1998, Iberall 1950). For the open path CO_2/H_2O instrument, which is mounted 0.25 m below the sonic, Kristensen's et al. (1997) results suggest that correction term is about +0.2%, which is negligible. However, at GLEES the pressure probe is mounted 0.33 m above the sonic. To date there are no models of covariance loss to the pressure covariance, $\overline{p'w'}$, resulting from vertical displacement. It is, therefore, possible that measured values of $\overline{p'w'}$ are underestimated because we neglect this aspect of system design. We do not include any possible influences the hose may have on the pressure fluctuations measured with the differential pressure sensor because again we expect this to be negligible (e. g., Holman 2001, Chapter 7). With these few exceptions, we use all the system design distances and the other instrument and block averaging filtering effects and Massman's (2000) analytical approach to estimate the system's equivalent time constant for high frequency, τ_e, and low frequency, τ_b, attenuation.

All GLEES turbulence data are sampled at 20 Hz and the flux averaging period is half an hour. Before any fluxes are calculated the 20 Hz data are screened for spikes using Højstrup's (1993) despiking and interpolating routine. The data are also tested for physically unrealistic values of skewness and kurtosis (Vickers and Mahrt 1997). Nevertheless, this specific period of time, consisting of 180 contiguous half hour periods, was chosen for this analysis because the data capture rate was extremely high and the data system and the data itself showed no significant problems. After despiking, the data were then used to form half-hourly mean wind speeds which are used with the planar fit method (Wilczak et al. 2001; Chapter 3) to define the coordinate system. Next, all 20 Hz data are rotated into this coordinate system and it is this rotated high frequency turbulence data that are used for calculating cospectra. For this study the GLEES cospectral data include the momentum covariance, $\overline{u'w'}$, the pressure covariance, $\overline{p'w'}$, the sonic virtual temperature covariance, $\overline{w'T'_v}$, and the water vapor $(\overline{w'\rho'_v})$ and CO_2 $(\overline{w'c'})$ covariances. In the case of the last two covariances we do not include the WPL terms (Webb et al. 1980) because we are correcting only the covariance between the sonic and the open-path CO_2/H_2O sensor.

3.2 Observed values of the cospectral parameters f_x, Δf_x, α, and $\Delta\alpha$

After rotation there are six steps that are performed to derive estimates of f_x and Δf_x.

- Each time series is padded with about 860 zeroes to make a total of 36,864 ($N_{FFT} = 2^{12}3^2$) data points. We note here that these winter time series showed no significant temporal half-hourly trends so that padding with zeroes should cause only a minimum of distortion in the cospectra (see Chapter 7 of Kaimal and Finnigan 1994). The FFT we employed here is very general and because it can accept any number of data points and it is not restricted to a power of 2. However, this FFT performs best and as fast as a standard FFT if the number of input data points is rich in powers of small primes. For a half-hour time series, sampled at 20 Hz, this value of N_{FFT} is the optimal value for FFT performance.

- After transforming the time series and forming the cospectra, the discrete cospectral estimates are corrected for the reduction associated with the padding by zeroes (Kaimal and Finnigan 1994).

- The raw cospectra are then smoothed with a frequency window that expands with frequency (Kaimal and Finnigan 1994). However, no other windowing or tapering was used on the time series or the cospectra (e. g., Kaimal and Finnigan 1994).

- The FFT frequencies, f, are nondimensionalized (multiplied) by the ratio z_{ref}/u to yield the nondimensional frequency, $\eta = fz_{ref}/u$.

- The smoothed cospectral estimates are then multiplied by frequency to form the frequency-weighted cospectra, which is fitted to the universal cospectra given by Equation 4.2 using the Levenberg-Marquardt nonlinear least squares algorithm (Press et al. 1992). We fit the frequency weighted cospectrum rather than the cospectrum itself because the frequency weighted cospectrum has more structure (i. e., it is not monotonic, which the cospectrum tends to be) so it provides a more reliable and precise estimate of the parameters. We also only fit the central portion of the observed cospectra to avoid influencing the fit by any high or low frequency attenuation. Note here that the transformation to nondimensional frequencies has no impact on Equation 4.2 or the results of the fitting procedure because $f/f_x = \eta/\eta_x$. Nor does it have any impact on the relative uncertainty in the correction factor, $\Delta F/F$ given by Equation 4.6, because $\Delta f_x/f_x = \Delta \eta_x/\eta_x$. Nevertheless, when presenting the f_x results we also include another nondimensional form for f_x, i. e., $\eta'_x = f_x(z_{ref} - d)/u$. This second formulation, which is more conventional, allows us to compare our results with those of previous studies and is used specifically in Equations 4.6 and 4.3 when parameterizing f_x. But because the displacement height, d, is somewhat uncertain, it also follows that η'_x is less certain that η_x. However, we do not include this source of uncertainty in our results.

- It is necessary to explore the parameter space in order to determine the optimal values of the parameters f_x, μ, and m, the slope parameter. Although the slope parameter is not the primary objective of the study it is useful to be able to examine the slope of the inertial subrange for its own interest as well as to assess its possible influence on the uncertainties in f_x and μ. Exploring the parameter space consists of fitting each of the cospectra several different times using different choices for the initial values of the parameters and different combinations of fixed and free parameters. The goal is to find a set of initial values for the parameters that produce the highest ensemble R^2 values (best fit to the ensemble of cospectra). This results in a set of optimized parameter values for each cospectrum. The ensemble mean and the root mean square variance of these optimized parameter values are then taken to be the final (best) values for a parameter and its inherent uncertainty or variability. For these fitting runs $m = 3/4$ is fixed. After determining the best estimates of η_x and μ, m is then varied between 0.7 and 0.8 to test the influence different m values may have on the quality of the fits. A final run is then performed where η_x, μ, and m are fit simultaneously to confirm that all parameter values are near optimal. This last run is performed primarily to confirm that $m \equiv 3/4$ (-7/3 cospectral decay) is reasonable for these cospectra.

The results of the tests concerning the slope parameter, m, confirmed that the inertial subrange decays according to -7/3 law and that no significant improvement in the quality of the fits could be achieved by another value of the -7/3 slope ($m = 3/4$). This was true for all five instrument covariances measured at GLEES.

The final values for μ, η_x, and η'_x and their associated uncertainties are listed in Table 4.1. M21's broadness parameter value is also listed in Table 4.1, but before discussing it is worthwhile comparing the present results with the results of Kaimal et al. (1972). For flat terrain and a neutral stability Kaimal et al. (1972) suggest that $\eta'_x = 0.085$ for the momentum flux, $\overline{u'w'}$, and 0.079 for the heat flux, $\overline{w'T'}$. This study indicates that the peak in the frequency weighted cospectrum is shifted to slightly lower values at GLEES for momentum flux and slightly higher values for the heat flux. But, the uncertainties are relatively large for the GLEES site, so an precise comparison is difficult.

From the formulations given by Kaimal and Finnigan (1994) for near neutral momentum cospectra, it is straightforward to obtain that the flat terrain cospectral broadness is $\mu = 1/2$. At the more topographically complex GLEES site, however, $\mu = 1$ indicating that the frequency-weighted cospectrum is more narrowly peaked than for flat terrain. For heat flux the comparison is complicated by the fact that the model cospectra presented by Kaimal and Finnigan (1994) is comprised of two equations (i. e., it is piecewise continuous). But the same general conclusion seems to hold for the broadness of heat flux cospectra as for momentum cospectra, except that the difference between the flat terrain and the GLEES heat flux cospectral broadness parameters, μ, is somewhat smaller and less significant than for the momentum broadness parameter. We conclude that, except for some subtle nuances, the momentum and heat cospectra over the forested complex terrain at GLEES are similar to the flat terrain momentum and heat cospectra. Furthermore, it is not clear how significant, if at all, these subtle differences are. This is because (i) Kaimal et al. (1972) did not use Equation 4.2 to describe their cospectra, (ii) they did not use the same coordinate system as the

Table 4.1. Optimal parameter values for instrument covariances measured at GLEES in Rocky Mountains of southern Wyoming during January 2001. At this site during wintertime the atmosphere is neutrally stable. The uncertainty associated with the value of each parameter is enclosed by parentheses. Here $\eta_x = f_x z_{ref}/u$ and $\eta'_x = f_x(z_{ref} - d)/u$.

Parameter	$\overline{p'w'}$	$\overline{u'w'}$	$\overline{w'T'_v}$	$\overline{w'\rho'_v}$	$\overline{w'c'}$
μ	0.25 (0.06)	1.0 (0.35)	0.6 (0.28)	0.6 (0.28)	0.6 (0.38)
η_x	0.22 (0.09)	0.14 (0.04)	0.21 (0.06)	0.19 (0.06)	0.22 (0.09)
η'_x	0.11 (0.05)	0.07 (0.02)	0.11 (0.03)	0.10 (0.03)	0.11 (0.05)
α	0.8 (0.2)	1.0 (0.2)	1.0 (0.2)	1.0 (0.2)	1.0 (0.2)

present study, and (*iii*) the variability we found in the GLEES cospectral parameter values suggests that Kaimal's et al. (1972) sample size may not have been adequate to determine the natural variability that was present at the time their data were obtained. Nevertheless, when using M21's analytical method to estimate the spectral correction factors appropriate to GLEES we will use the data from Table 4.1, rather than the cospectral model from Kaimal et al. (1972).

Table 4.1 also indicates that the measured water vapor and CO_2 covariance cospectra largely follow the heat covariance cospectra because without the WPL terms these covariances are dominated by the heat flux term. In fact the CO_2 covariance is probably more strongly influenced by the heat flux term than is the water vapor covariance because the true CO_2 flux is quite small, whereas the true water vapor flux can be quite high even during the winter. Therefore, the heat flux term has a relatively smaller contribution to the $\overline{w'\rho'_v}$ covariance than to the CO_2 covariance. This and the possibility of heat and vapor dissimilarity may explain why the cospectral parameter value of η_x or η'_x for water vapor differ more from the heat covariance values than do the CO_2 covariances. We also note that the uncertainties in the parameter values for the CO_2 covariance are greater than for either $\overline{w'T'_v}$ or $\overline{w'\rho'_v}$. This may result from the fact that the CO_2 covariance combines the variability in all the other fluxes, CO_2, water vapor, and heat.

Finally, because Kaimal et al. (1972) did not measure the pressure covariance, $\overline{p'w'}$, it is not possible to compare the present results with theirs. However, Table 4.1 does indicate that the $\overline{p'w'}$ cospectra are considerably broader than $\overline{u'w'}$ cospectra. This may be a general characteristic of $\overline{p'w'}$ cospectra as Wilczak (personal communication) has found the same characteristic also occurs over the ocean.

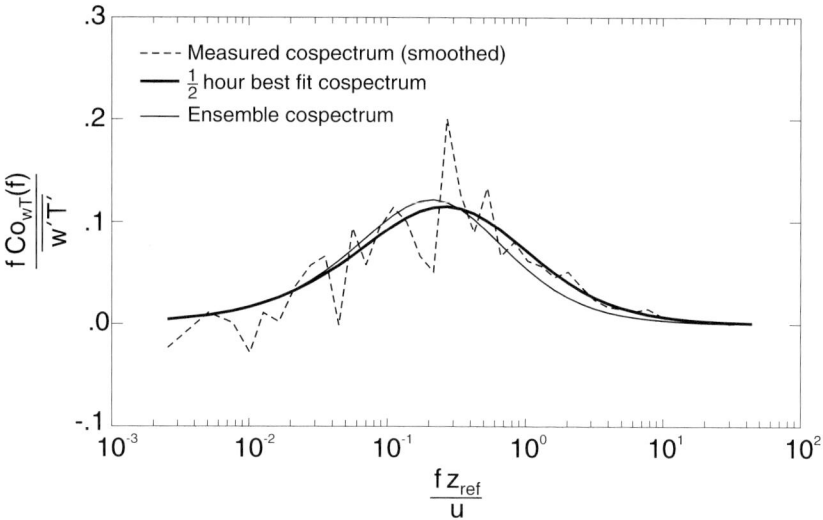

Figure 4.3. Comparison of the smoothed FFT cospectrum for $\overline{w'T_v'}$ obtained at 6:30 am MST January 2, 2001 at the GLEES (dashed line) with the optimal fit to this cospectrum with Equation 4.2 (solid line) and the modeled cospectrum derived from Equation 4.2 using the ensemble best fit parameters given in Table 4.1 (dotted line).

The parameter α and its associated uncertainty, $\Delta\alpha$, are determined empirically by matching the correction factors generated by M21 analytical approach to those generated by direct integration using Equations 4.1 and 4.2. The results, shown in Table 4.1, indicate, as M21 claimed, that the analytical method is not particularly sensitive to the broadness parameter. The only possible exception to this in this study are the $\overline{p'w'}$ cospectra ($\alpha = 0.8$), which are extraordinarily broad ($\mu = 0.25$) when compared to the other cospectra ($\mu > 0.50$). M21 also found that Kaimal's et al. (1972) flat terrain cospectra for neutral or unstable atmospheric stability also showed sufficient broadness that $\alpha = 0.925$ was required for a good match. These two non-unity values of α suggest that $\Delta\alpha$ can be estimated from the observed range of variability in the different computed values for α. We, therefore, take $\Delta\alpha = 0.2$ since the observed values of α range from 0.8 to 1.0. For the purposes of estimating $\Delta F/F$ from Equation 4.6 we will take the nominal value of α as 1, as was discussed previously.

As an example of the different steps involved in this analysis, Figure 4.3 compares the three following cospectra: the smoothed FFT cospectrum for $\overline{w'T_v'}$ obtained at 6:30 am MST January 2, 2001 (dashed line), the optimal fit of the FFT cospectrum by Equation 4.2 (solid line), and the modeled cospectrum derived from Equation 4.2 using the ensemble best fit parameters given in Table 4.1 (dotted line). This figure clearly indicates the nature of the three types of cospectral variability mentioned previously.

3.3 Cospectral (dis)similarity between sites: A CarboEurope flux site

The last section includes a brief comparison between the flat terrain (momentum and heat) cospectra of Kaimal et al. (1972) and the cospectra obtained at a topographically complex site during neutral atmospheric conditions. The results indicated that there were cospectral differences between these sites, but it was less clear how significant the differences might be for discussions of cospectral similarity. For spectral attenuation issues the differences are quite significant because η'_x for momentum flux was about 18% lower at GLEES than at the flat terrain site, whereas η'_x for heat flux was almost 40% higher at GLEES. It is, therefore, worthwhile to include another site in this comparison in an effort to determine how generalizable the previous results might be from one site to another.

This section presents values of η'_x were obtained at Griffin Forest, a CarboEurope flux site located at [56°36'30"N, 3°47'15"W] near Aberfeldy, Perthshire, Scotland. The forest in this region covers about 75% of the surface and is 97.3% Sitka spruce (*Picea sitchensis*), 2.1% Douglas fir (*Pseudotsuga menziesii*) and 0.6% birch (*Betula pendula*). The remaining 25% of the surface is open space: roads, streams, and ridges dominated by heather and grasses. The forest leaf area index (LAI) is about 8, and where the canopy is closed, the understory is quite sparse and dominated by mosses. The canopy height is about 7.5 m and wind speed profile measurements suggest that $d = 5.3$ m or $\approx 0.7h$.

The site has an elevation of 350 m ASL, and is also a micrometeorologically complex site, being located in a region of ridge and valley ranging in height between 50 and 600 m. It is situated in a broad northwest facing valley with a mean slope of 9%. The fetch to the southwest is about 500 m before it is disturbed by a hill. To the northwest the fetch is about 2000 m and to the northeast and southeast it is about 1000 m. The eddy covariance instruments are mounted at 15.5 m and consist of a Solent R2 three dimensional sonic, a closed-path Licor (6262) CO_2/H_2O infrared gas analyzer (IRGA), and an open-path Licor 7500 IRGA. The inlet to the closed-path IRGA is located 0.20 m below and 0.05 m to the north of the sonic. The intake tube length is 18 m and it has an inside diameter of 0.00623 m. The flow rate is 6 L min^{-1} and the flow

Reynolds number is about 1400. The open-path sensor is positioned 0.30 m from the center of the sonic. Data from this CarboEurope flux site were obtained between July and September 2000. They were collected at 20.833 Hz ($N_{1/2} = 37500$ data points per half-hour) and were processed by the Edisol/EdiRe software package. The eddy correlation system and the data processing package are described in more detail by Moncrieff et al. (1997). All covariances are calculated in the planar fit coordinate system (Wilczak et al. 2001; Chapter 3).

All Griffin forest cospectra are calculated for each 2-hour period using smoothed and spliced FFT (Kaimal and Finnigan 1994). To obtain the high frequency portion of the cospectra the 20.833 Hz time series are divided into M sequential segments of contiguous 512-point time series. [Note that for this method of cospectral analysis $N_{FFT} = 2^9$ and $M = 4N_{1/2}/N_{FFT}$ if N_{FFT} divides $4N_{1/2}$ exactly, otherwise $M = 1+$ the integer part of $\{4N_{1/2}/N_{FFT}\}$.] If necessary, the final segment is zero-buffered to insure that N_{FFT} is the same for all M data segments. Each segment is detrended using a linear least squares fit and tapered using a Hamming window (Kaimal and Kristensen 1991). The FFT is applied to each segment and the transformed results are averaged prior to logarithmic bin-averaging of the spectral estimates. The low frequency cospectrum for the same data period is obtained by averaging M sequential data points to obtain a single set of N_{FFT} averaged data points. Again this data series is detrended, tapered, transformed, and logarithmically bin-averaged using the same methods as with the high frequency portion. The resulting high and low frequency portions are merged and sorted by frequency to obtain a single cospectral curve, to which Equation 4.2 is then fitted. About 200 cospectra for both stable and neutral/unstable atmospheric conditions were available for analysis.

Table 4.2 presents the cospectral parameter values for η'_x obtained for Griffin forest. The other cospectral parameter values are not included because they are of less significance. There are several key observations concerning the Griffin forest results that need to be emphasized. First, comparing the neutral stability η'_x values with those of GLEES (Table 4.1) indicates that for neutral conditions η'_x at Griffin forest are associated with much lower frequencies than at GLEES. Second, η'_x at Griffin forest tends to decrease with increasing atmospheric instability and that η'_x for stable conditions tend to be relatively constant and highly variable. This stability dependency is opposite of the flat terrain site of Kaimal et al. (1972), where constant η'_x is associated with unstable conditions and η'_x tends to increase with increasing atmospheric stability. Third, the open-path water vapor and CO_2 covariances and

Table 4.2. Optimal values of η'_x for instrument covariances measured at Griffin forest, Scotland, during July through September 2000. The uncertainty associated with the value of each parameter is enclosed by parentheses. Here $\eta'_x = f_x(z_{ref} - d)/u$.

Covariance	η'_x very unstable $z/L \leq -2$	η'_x neutral stability $-0.2 < z/L < 0$	η'_x stable $z/L > 0$
$\overline{u'w'}$	0.009 (0.063)	0.027 (0.063)	0.029 (0.015)
$\overline{w'T'_a}$	0.010 (0.032)	0.073 (0.032)	0.043 (0.025)
$\overline{w'\rho'_v}$ (closed-path)	0.003 (0.008)	0.015 (0.008)	0.014 (0.011)
$\overline{w'c'}$ (closed-path)	0.011 (0.015)	0.043 (0.015)	0.041 (0.023)
$\overline{w'\rho'_v}$ (open-path)	0.014 (0.019)	0.041 (0.019)	0.036 (0.024)
$\overline{w'c'}$ (open-path)	0.009 (0.011)	0.047 (0.011)	0.040 (0.021)

the closed-path CO_2 covariance can probably be considered cospectrally similar for any atmospheric stability, especially given the uncertainty of their respective η'_x values. But, some divergence between open- and closed-path results should not be unexpected because the influence of the WPL temperature covariance (heat flux) term will be considerably greater on the open-path covariance than on the closed-path covariance. However, the closed-path water vapor cospectra are so different from these other three scalar cospectra that they are probably untrustworthy. We hypothesize that the water vapor fluctuations have been so strongly attenuated by the tube flow and interactions with the flow path walls that their associated η'_x values should not be used for spectral attenuation issues. Fourth, the $\overline{w'T'_a}$ cospectra is for the atmospheric heat flux not for the sonic temperature covariance; but, any differences introduced by using $\overline{w'T'_a}$ rather than $\overline{w'T'_v}$ should be small. Fifth, and final, no continuity requirement between stable and neutral values of η'_x was imposed. Except for $\overline{w'T'_a}$, the results suggest that the η'_x values are probably continuous across these two stability classes. Although the variability of η'_x for $\overline{w'T'_a}$ is high in both stability classes, the data suggest that the value of 0.073 for the neutral case is likely to be biased high.

In summary, the results from this CarboEurope flux site, when combined with those from GLEES and Kaimal et al. (1972), suggest that cospectra, which use the η'_x normalization to parameterize f_x, possess considerable site-to-site variablity. Therefore, any investigation of issues involving spectral attenuation of eddy covariance data needs to be examined on a site specific basis and that the arbitrary use of Kaimal's et

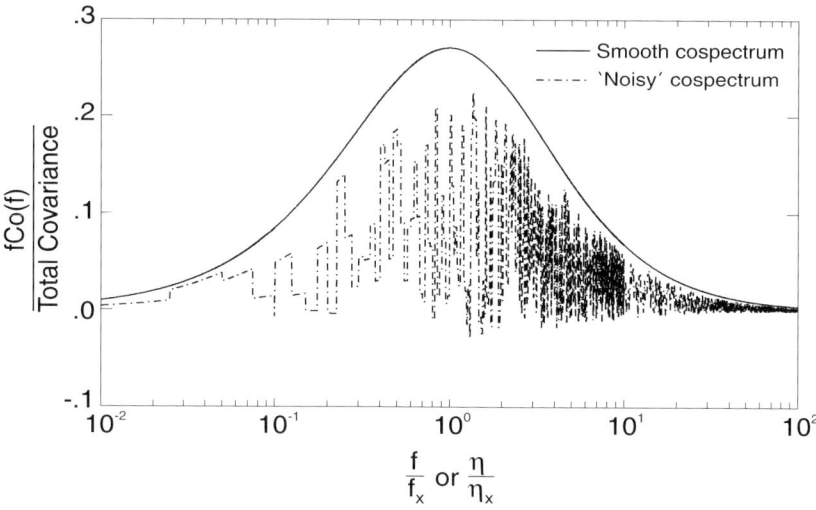

Figure 4.4. A comparison of the hypothetical frequency-weighted atmospheric cospectrum shown in Figure 4.1 (solid line) with a version of it after modification by a random number generator (dash-dot). This method of modifying the smooth cospectral shape is used to estimate the uncertainty in the spectral correction factor that is associated with departures of true cospectra from smoothed versions of the cospectra.

al. (1972) model for η'_x has the potential to introduce significant errors into spectral correction factors for eddy covariance fluxes.

Besides any possible site-to-site variability in cospectra there is also the inherent variability of cospectra from one flux-averaging period to the next.

3.4 Departures from smooth cospectral shapes

As Figure 4.3 clearly indicates observed cospectra are not particularly smooth. This departure from a smooth cospectral shape is another source of uncertainty in spectral corrections. Although obviously related to possible uncertainties in f_x and μ, this source of uncertainty is, nevertheless, different than these other two. The ideal method for spectral corrections might very well be to use the transfer function method, Equation 4.1, combined with half hourly cospectra, which would thereby include the half-hourly departures from a smooth cospectral shape. While this is difficult to achieve, it can be emulated by modifying the smooth cospectral shapes given by Equation 4.2 using a random number generator (Press et al. 1992) to change the smooth cospectra into a much 'noisier' cospectra than would traditionally be used. Figure 4.4 compares a smooth cospectra with a noisier cospectra produced by this method. By providing several different integer seed values (Press et al. 1992) to the random number generator and calculating the spectral correction factor with many different noisy cospectra, we quantify the effects of random variablity in the cospectra on the spectral correction factor. Each of the spectral correction factors developed with a non-smooth cospectra using the integral method is compared to the correction factors determined by M21's analytical method, which uses the smooth cospectra. This produces a range of variation in F that can be conveniently incorporated into Equation 4.6 for $\Delta F/F$ by simply multiplying Equation 4.6 by a constant, $C = 1.2$. Note that the value of 1.2 was synthesized from all five of the GLEES covariance measurements. A separate determination of C for a closed path CO_2 system was also performed and is discussed in section 3.5. The final expression for the relative uncertainty, $\Delta F/F$, is

$$\frac{\Delta F}{F} = C\{[(1+\frac{1}{b})p - \frac{1+p}{b}]^2[\frac{\Delta f_x}{f_x}]^2$$
$$+[(1+\frac{1}{b})p\ln(p) - \frac{1+p}{b}\ln(b)]^2[\Delta\alpha]^2$$
$$+[(1+\frac{1}{b})p]^2[\frac{\Delta\tau_e}{\tau_e}]^2\}^{0.5}/[(1+p)(1+\frac{1}{b})] \qquad (4.7)$$

3.5 Results for an open-path system

Figure 4.5 shows relative uncertainty in the spectral correction factor, $\Delta F/F$, as a function of wind speed for the GLEES CO_2 covariance measurement. This figure suggests that applying a spectral correction factor based on the transfer function approach to GLEES wintertime CO_2 covariances would lead to a 5 to 10% uncertainty in the resulting CO_2 covariance at low wind speeds and a 2 to 4% uncertainty at higher wind speeds. Figure 4.6 shows the correction factor, F, as a function of wind speed for the GLEES CO_2 covariance measurement as estimated by the analytical method of M21 along its inherent range of uncertainty as predicted by Equation 4.7. The filters and equivalent time constants employed for this and the previous figure are taken from Table 1 of Massman (2000) and include sonic line averaging for scalar fluxes, lateral separation, longitudinal separation without a first order instrument, and high pass block averaging. The results shown in this figure indicate that the uncertainty associated with a relatively larger value of F is itself relatively large and that the inherent uncertainty in F decreases as F decreases. This supports the conclusion that if an eddy covariance

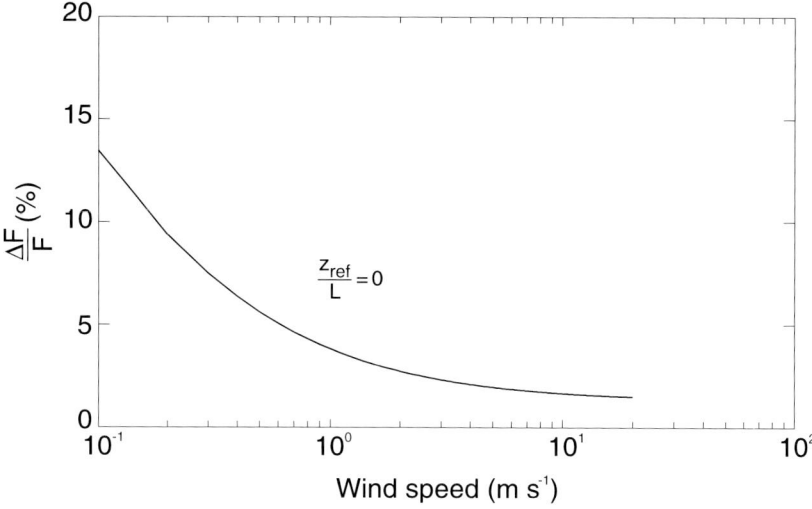

Figure 4.5. The relative uncertainty in the spectral correction factor, $\Delta F/F$, as a function of wind speed for the GLEES CO_2 covariance measurement. Neutral atmospheric stability is assumed.

system is designed so that spectral attenuation is minimal (and therefore relatively small), then the transfer method of estimating spectral correction factors is not particularly sensitive to the shape, the lack of smoothness in the cospectra, or the natural variablity of the cospectra. To rephrase, if the spectral correction factors are small, then the transfer function method should provide trustworthy and relatively accurate values for the spectral correction factors.

3.6 Extension to a closed-path system and $\Delta \tau_e$

The GLEES data provide an example of an eddy covariance system with an open path CO_2 sensor. Now we wish to examine the same system with a closed path sensor. We construct this scenario primarily to test our present methods on a system that has a much longer equivalent time constant for high frequency attenuation, τ_e, than the open path system. All the methods and calculations remain the same, except we assume that the closed path system has an intake tube attached to a first order instrument, which is housed in a separate shelter at the base of the tower. We further assume that the time constant for the first order instrument (τ_1) was determined in situ to be 0.1 s and we will take $\Delta \tau_e = 25\ \%$ of τ_1. These values for τ_1 and $\Delta \tau_e$ are to be understood as purely hypothetical. For any real application they could depart signifi-

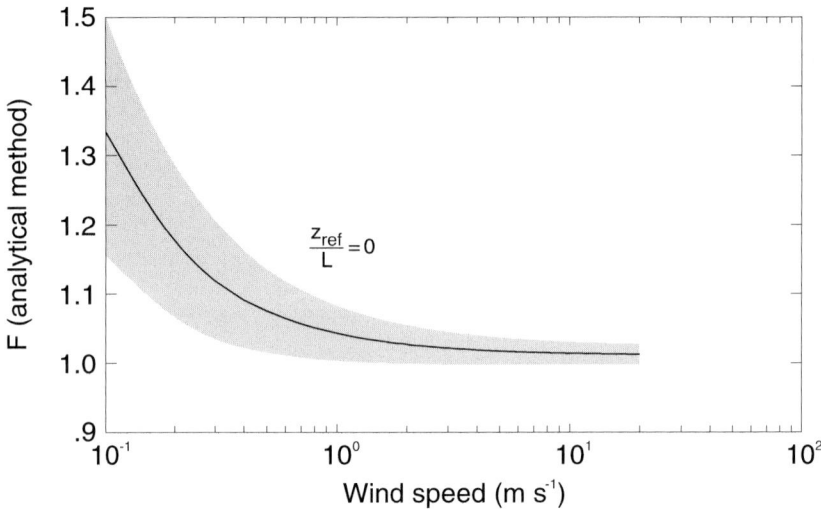

Figure 4.6. The correction factor, F, as a function of wind speed for the GLEES CO_2 covariance measurement as estimated by the analytical method of M21 (dark solid line) and its inherent range of uncertainty as predicted by Equation 4.7 (shaded area). The factor C used with Equation 4.7 to account for variable (random) departures from a smooth cospectral shape is 1.2.

cantly from these present values and from one eddy covariance system to another.

In this simulation we allow for the attenuation associated with tube flow (M21; Massman 1991), a small amount of lateral separation, some small amount of longitudinal separation with first order effects (Massman 2000), and a first order time constant for the CO_2 sensor equal to 0.1 s. Because this closed path scenario falls into M21's category for systems characterized by equivalent time constants between 0.1 and 0.3 s and a neutrally stable atmosphere, his analytical approach would suggest that Equation 4.3 above be augmented by the term $[1 + 0.9p^\alpha]/[1 + p^\alpha]$. Here we do not employ the additional term. Rather we use Equation 4.3 directly. Figure 4.7 shows the difference between the integral method, Equation 4.1, and Equation 4.3. For all wind speeds the difference is small and the agreement is quite good.

Figure 4.8 shows the correction factor, F, as a function of wind speed for the simulated closed path CO_2 covariance measurement as estimated with analytical method of M21 along its inherent range of uncertainty as predicted by Equation 4.7. For this example, the factor of C in Equation 4.7 was again found to be 1.2 using the same method as the open path sensor. This figure is the closed path analog to Figure 4.6

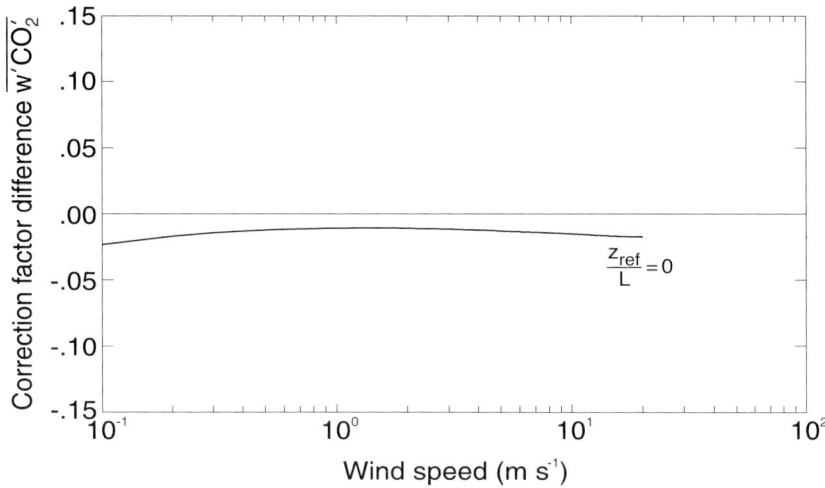

Figure 4.7. Difference between the spectral correction factors estimated from the numerical integration of Equation 4.1 and Equation 4.3 from M21. These calculations are for a closed-path eddy covariance system and a neutral atmosphere and are shown as a function of the horizontal wind speed. The zero line is highlighted.

and clearly demonstrates the same conclusions. That is, if the correction factors are relatively large, which they are here by design, then the inherent uncertainties associated with the transfer function method can also be relatively large. For example the calculations shown in Figure 4.8 suggest that for wind speeds between about 2 and 10 m s^{-1} the correction factor is about $1.1 \pm 10\%$. For this same range of wind speeds Figure 4.6 suggests that the open path system has a correction factor and an inherent uncertainty about half of that. In these two examples the fundamental difference between the open and closed path systems is their respective time constants, reinforcing the notion that longer time constants result in more spectral attenuation, greater uncertainty in the spectral corrections, and, therefore, greater inherent uncertainty in the fluxes or covariances that are spectrally corrected with the transfer function method.

Most of the uncertainty shown in Figure 4.8 results from variability in the cospectra and from the uncertainty in the cospectral parameters, not from $\Delta\tau_e$. This is because the present example is fairly conservative in its estimate of τ_1 and $\Delta\tau_e$. As either of these parameters increase in value so also do the estimates for F and ΔF. At some eddy covariance sites τ_1 can exceed 1 s, which is considerably longer than the present hypothetical value of 0.1 s. At present there has been no information

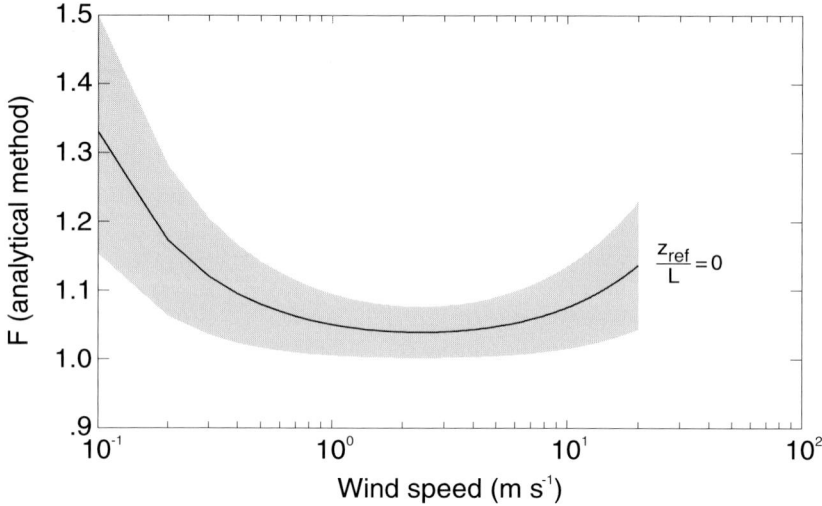

Figure 4.8. The correction factor, F, as a function of wind speed for the GLEES CO_2 covariance measurement as estimated by the analytical method of M21 (dark solid line) and its inherent range of variablity as predicted by Equation 4.2 (shaded area). The factor C used with Equation 4.2 to account for the variable (random) departures from smooth cospectral shape is 1.2, which is the same as with the GLEES open path sensor.

published (of which we are aware) that discusses $\Delta\tau_1$, which we would suggest be used to estimate $\Delta\tau_e$. However, we can easily imagine that $\Delta\tau_1$ can exceed the value of 0.025 s we use in this study.

3.7 Discussion and caveats concerning low frequencies

One of the advantages of the analytical method is the ability to estimate the high and low frequency portions of the correction factors separately. Rewriting Equation 4.4 using the b and p notation yields $F = [1 + 1/b][1 + p]$, where $\alpha = 1$ is assumed for simplicity. The high frequency correction factor is expressed by the p term, $[1 + p]$, while the low frequency portion is $[1 + 1/b]$. Table 4.3 (using η'_x from Table 4.1) lists these two terms for the GLEES covariance data and the simulated closed path CO_2 data. For the open path covariances both corrections are small; however, the low frequency portion tends to be slightly larger than the high frequency portion. In the case of the closed path CO_2 system the high frequency portion is by far the dominant term, which indicates the influence of the relatively longer high frequency time constant, τ_e, for closed path system than for the open path.

Table 4.3. Simplified (and approximate) high and flow frequency correction factors for the GLEES eddy covariance system estimated with M21's analytical method. The $(1+p)$ factor is for high frequency effects and the $(1+1/b)$ factor is for low frequency effects. Wind speeds are assumed to be greater than $1~\mathrm{m\,s^{-1}}$.

Covariance	$(1+p)$	$(1+1/b)$
$\overline{p'w'}$	1.04	1.01-1.06
$\overline{u'w'}$	1.00-1.01	1.00-1.05
$\overline{w'T_v'}$	1.00-1.01	1.00-1.03
$\overline{w'\rho_v'}$	1.01	1.00-1.04
$\overline{w'c'}$ (open)	1.01	1.00-1.03
$\overline{w'c'}$ (closed)	1.02-1.13	1.00-1.03

Nevertheless, both high and low frequency correction factors are based on extrapolations into regions of the cospectra where there are no direct observations. In the case of the high frequency corrections, this extrapolation is probably quite trustworthy because the -7/3 decay law is theoretically sound and has been observationally verified in many studies. But, considerably less is known about the very low frequency portion of the cospectra. For any sampling period, T_b, there is no observationally based information on low frequencies within the spectral band $[0, 1/T_b]$. Therefore, by including the high pass filters (block averaging filters, etc.) in Equation 4.1 and integrating over the whole spectral waveband $[0,\infty]$ we implicitly assume that the observed turbulence cospectra extends smoothly and continuously into the unobserved portion of the cospectra. This assumption will not be true for all atmospheric conditions, like those that might occur during convective conditions or when other large scale planetary boundary layer processes are active at the time of the measurements. Under these conditions, the low frequency correction factor, $[1 + 1/b]$, will likely misrepresent the true correction factor.

At GLEES during the wintertime such large scale planetary boundary layer effects do not seem to be significant. For example, Figure 4.9 is an ogive computed from one of the 180 half hourly periods for the temperature covariance $\overline{w'T_v'}$. This particular ogive indicates that virtually all the half hourly cospectral power is contained in motions with periods shorter than about 5.5 minutes because the cumulative cospectral power has reached a relatively stable maximum at 0.003 Hz. We should note here that this particular example is an extreme case. Other ogives (not shown) indicate that motions somewhat longer than 30 minutes can contribute to half hourly at GLEES during the wintertime. However, we also performed other tests on the data by concatenating two

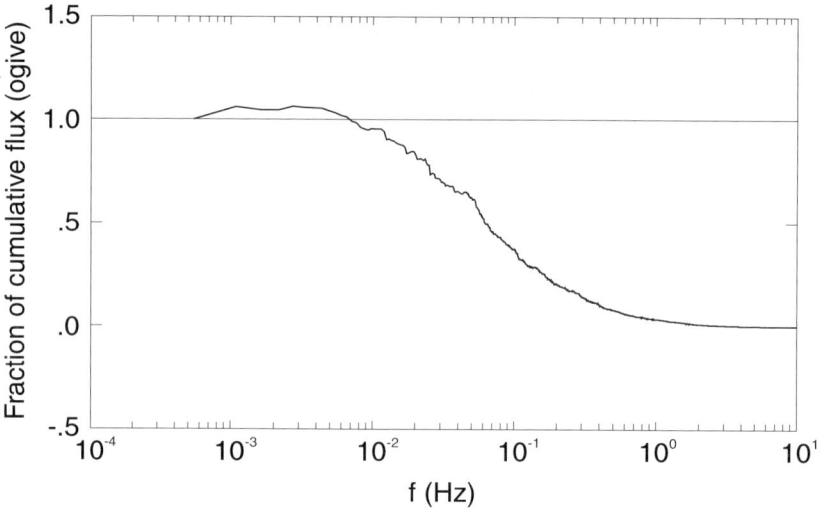

Figure 4.9. An ogive calculated from the unsmoothed FFT cospectrum for the temperature covariance $\overline{w'T_v'}$. These data were obtained at 6:30 am MST January 2, 2001 at the GLEES. Neutral atmospheric stability is assumed. This example indicates that virtually all cospectral power is located in atmospheric motions with periods less than about 5.5 minutes. Other examples (not shown) show contributions from longer period motions. The 1.0 line is highlighted.

half hourly time series to form a single hour time series and by subsampling and concatenating four half hourly time series to form a two hour time series. We could find no evidence of any long period, low frequency, flux contributions. Of course, this result should not be unexpected because high elevation, wintertime, high-wind, high-turbulence conditions are not particularly good candidates for low frequency planetary boundary layer motions. But when conditions are more conducive these slow large scale motions will contribute to the fluxes (Sakai et al. 2001, Finnigan et al. 2003). At present there is no single agreed-upon well defined method for correcting fluxes for these low frequency contributions. Consequently, the spectral transfer method as it is posed by Equations 4.1 and 4.2 should not be understood as compensating for all flux loss due to undersampling the low frequency planetary boundary layer motions. Issues concerning the influence of these types of motions on measured fluxes is discussed in Chapter 5 of this volume.

3.8 Summary

This study has developed a method of estimating the uncertainty in any measured covariance that is spectrally corrected by the transfer function method. It includes the effects of the (flux-averaging) period-to-period variability in (*i*) the frequency-weighted cospectral peak, f_x or η'_x, (*ii*) the random departure of measured cospectra from a smooth shape (encapsuled by the parameter C of Equation 4.7), (*iii*) the broadness of the frequency-weighted cospectra (μ or α), and (*iv*) some of the possible uncertainties in the effective time constant of an eddy covariance system, τ_e. This model, summarized by Equation 4.7, was calibrated mostly using cospectral data from one particular AmeriFlux site (GLEES), but also included comparisons with similar data from Kaimal et al. (1972) and an CarboEurope flux site (Griffin forest). Conclusions reached in this study are

- The uncertainty associated with variability in the cospectral broadness is constant, i. e., $\Delta\alpha = 0.2$.

- It is reasonable to simplify Equation 4.7 by approximating relative uncertainty in f_x, i. e., $\Delta f_x / f_x$, by 0.4 or 0.5; where Δf_x is a measure of the inherent variability in f_x. The value of 0.4 is the upper bound of values found at GLEES (Table 4.1) and 0.5 is a value more representative of the Griffin forest site (Table 4.2). This simplification eliminates the need to evaluate Δf_x directly.

- Spectral corrections and their associated uncertainties are likely to be site specific because the nondimensional frequency, $\eta'_x = f_x(z_{ref} - d)/u$, was shown to be site specific in this study. In fact, this study has shown that η'_x varies significantly between the flat terrain site of Kaimal et al. (1972) and the micrometeorologically complex forested sites of GLEES and Griffin forest. Therefore, there is no a priori reason to assume that the values for nondimensional frequency, η'_x, developed in this study or by Kaimal et al. (1972) apply universally. In turn this suggests that to find any truly universal cospectral shape will require a different turbulent time scale than $(z_{ref} - d)/u$.

- Equation 4.7 uses a value of 1.2 for the parameter C, which seems to be reasonable for both open and closed path CO_2 systems. However, we did not emulate all possible or observed eddy covariance systems, so it is conceivable that C could be somewhat different for other eddy covariance systems. Thus there may remain a need to calibrate Equation 4.7 at more eddy covariance sites.

- The uncertainty in spectral correction, F, estimated with the transfer function method, ΔF, increases as F increases and decreases with decreasing F. Careful attention to minimizing separation distances and instrument time constants should help keep high frequency losses small so that F and ΔF are small. In this case the high frequency correction factors should be relatively independent of cospectral shape, lack of cospectral smoothness, and inherent variability.

- The spectral losses at low frequencies on the other hand are not so much instrument related as they are related to the length of the sampling period and the nature of the low frequency atmospheric motions present during a sampling period. Because so little is known about the nature of these low frequency atmospheric motions (1-4 hours) it is difficult to make specific recommendations for reducing this undersampling error. However, in general, as the measurement height decreases the low frequency content of any measured flux should also decrease, which will reduce the problem somewhat. Unfortunately, this improvement will be offset by increasing high frequency content in the fluxes.

- Present results indicate that further research is needed into the low frequency (1-4 hour) components of eddy covariance fluxes and in the nature of the differences in the frequency-weighted cospectral peaks, η'_x and f_x, between different sites.

4 References

Aubinet, M., Chermanne, B., Vandenhaute, M., Longdoz, B., Yernaux, M., Laitat, E.: 2001, 'Long term carbon dioxide exchange above a mixed forest in the Belgian Ardennes', *Agric. For. Meteorol.* **108**, 293-315.

Auble, D. L., Meyers, T. P.: 1992, 'An open path, fast response infrared absorption gas analyzer for H_2O and CO_2', *Bound.-Layer Meteorol.* **59**, 243-256.

Aydin, I.: 1998, 'Evaluation of fluctuating pressure measured with connection tubes', *J. Hydraulic Engineering*, **124**, 413-418.

Busch, N. E.: 1973, 'Turbulent transfer in the atmospheric surface layer', In: (Ed.) Haugen, DA, *Workshop on Micrometeorology*, American Meteorological Society, Boston, 1-28.

Coleman, H. W., Steele, W. G.: 1999, *Experimentation and Uncertainty Analysis for Engineers*. John Wiley, New York, 275 pp.

Cook, R. K., Bedard, A. J.: 1971, 'On the measurement of infrasound', *Geophysical Journal Royal Astronomical Society* **26**, 5-11.

Finnigan, J. J., Clement, R., Mahli, Y., Leuning, R., Cleugh, H. A.: 2003, 'A reevaluation of long-term flux measurement techniques: Part 1. Averaging and coordinate rotation', *Bound.-Layer Meteorol.* **107**, 1-48.

Goulden, M. L., Daube, B. C., Fan, S.-M., Sutton, D. J., Bazzaz, A., Munger, J. W., Wofsy, S. C.: 1997, 'Physiological response of a black spruce forest to weather', *J. Geophys. Res.* **102**, 28,987-28,996.

Gurvich, A. S.: 1962, 'The pulsation spectra of the vertical component of the wind velocity and their relations to micrometeorological conditions', *Izvestiya Atmospheric Oceanic Physics* **4**, 101-136.

Hicks, B. B.: 1972, 'Propellor anemometers as sensors of atmospheric turbulence', *Bound.-Layer Meteorol.* **3**, 214-228.

Hicks, B. B., McMillen, R. T.: 1988, 'On the measurement of dry deposition using imperfect sensors and in non-ideal terrain', *Bound.-Layer Meteorol.*, **42**, 79-84.

Højstrup, J.: 1993, 'A statistical data screening procedure', *Measurement Science Technology* **4**, 153-157.

Holman, J. P.: 2001, *Experimental Methods for Engineers*. 7th Edition. McGraw-Hill, Boston, MA, 698 pp.

Horst, T. W.: 1973, 'Spectral transfer functions for a three-component sonic anemometer', *J. Appl. Meteorol.* **12**, 1072-1075.

Horst, T. W.: 1997, 'A simple formula for attenuation of eddy fluxes measured with first-order-response scalar sensors', *Bound.-Layer Meteorol.* **82**, 219-233.

Horst, T. W.: 2000, 'On frequency response corrections for eddy covariance flux measurements', *Bound.-Layer Meteorol.* **94**, 517-520.

Horst, T. W., Oncley, S. P., Semmer, S. R.: 1997, 'Measurement of water vapor fluxes using capacitance RH sensors and cospectral similarity', In: *12th Symposium on Boundary Layers and Turbulence*, American Meteorological Society, Boston, 360-361.

Iberall, A. S.: 1950, 'Attenuation of oscillatory pressures in instrument lines', *Journal Research National Bureau Standards* **45**, 85-108.

Kaimal, J. C., Clifford, S. F., Lataitis, R. J.: 1989, 'Effect of finite sampling on atmospheric spectra', *Bound.-Layer Meteorol.* **47**, 337-347.

Kaimal, J. C., Finnigan, J. J.: 1994, *Atmospheric Boundary Layer Flows - Their Structure and Measurement*. Oxford University Press, New York, NY, 289 pp.

Kaimal, J. C., Kristensen, L.: 1991, 'Time series tapering for short data samples', *Bound.-Layer Meteorol.* **57**, 187-194.

Kaimal, J. C., Wyngaard, J. C., Haugen, D. A.: 1968, 'Spectral characteristics of surface-layer turbulence', *Quart. J. R. Meteorol. Soc.* **98**, 563-589.

Kaimal, J. C., Wyngaard, J. C., Izumi, Y., Coté, O. R.: 1972, 'Spectral characteristics of surface-layer turbulence', *Quart. J. R. Meteorol. Soc.* **98**, 563-589.

Katul, G. G., Hsieh, C. I.: 1999,'A note on the flux-variance similarity relationship for heat and water vapor in the unstable atmospheric surface layer', *Bound.-Layer Meteorol.* **90**, 327-338.

Koprov, B. M., Sokolov, D. Y.: 1973, 'Spatial correlation functions of velocity and temperature components in the surface layer of the atmosphere', *Izvestiya Atmospheric Oceanic Physics* **4**, 178-182.

Kristensen, L., Mann, J., Oncley, S. P., Wyngaard, J. C.: 1997, 'How close is close enough when measuring scalar fluxes with displaced sensors', *J. Atmos. Oceanic Technol.* **14**, 814-821.

Laubach, J., McNaughton, K. G.: 1999, 'A spectrum-independent procedure for correcting eddy fluxes measured with separated sensors', *Bound.-Layer Meteorol.* **89**, 445-467.

Laubach, J., Teichmann, U.: 1996, 'Measuring energy budget components by eddy correlation: Data corrections and application over low vegetation', *Beiträge Physik Atmosphäre*, **69**, 307-320.

Lee, X., Black, T. A.: 1994, 'Relating eddy correlation sensible heat flux to horizontal sensor separation in the unstable atmospheric surface layer', *J. Geophys. Res.* **99**, 18,545-18,553.

Massman, W. J.: 1991, 'The attenuation of concentration fluctuations in turbulent flow through a tube', *J. Geophys. Res.* **96**, 15,269-15,273.

Massman, W. J.,: 2000, 'A simple method for estimating frequency response corrections for eddy covariance systems', *Agric. For. Meteorol.* **104**, 185-198.

Massman, W. J.: 2001, 'Reply to comment by Rannik on "A simple method for estimating frequency response corrections for eddy covariance systems"', *Agric. For. Meteorol.* **107**, 247-251.

Massman, W. J., Lee, X.: 2002, 'Eddy covariance flux corrections and uncertainties in long-term studies of carbon and energy exchanges', *Agric. For. Meteorol.* **113**, 121-144.

Mestayer, P. G., Larsen, S. E., Fairall, C. W., Edson, J. B.: 1990, 'Turbulence sensor dynamic calibration using real-time spectral computations', *J. Atmos. Oceanic Technol.* **7**, 841-851.

Moncrieff, J., Massheder, J., Verhoef, A., Elbers, J., Heusinkveld, B., Scott, S. L., DeBruin, H. A. R., Kabat, P., Soegaard, H., Jarvis, P.: 1997, 'A system to measure Surface fluxes of momentum, sensible heat, water vapour and carbon dioxide', *J. Hydrol.* **188**, 589-611.

Moore, C. J.: 1986, 'Frequency response corrections for eddy correlation systems', *Bound.-Layer Meteorol.* **37**, 17-35.

Nishiyama, R. T., Bedard, A. J.: 1991, 'A "Quad-Disc" static pressure probe for measurement in adverse atmospheres: With a comparative review of static pressure probe designs', *Review Scientific Instruments* **62**, 2193-2204.

Press, W. H., Teutkolsky, S. A., Vetterling, W. T., Flannery, B. P.: 1992, *Numerical Recipes*, Second Edition, Cambridge University Press.

Rannik, Ü.: 2001, 'A comment on the paper by W.J. Massman "A simple method for estimating frequency response corrections for eddy covariance systems"' *Agric. For. Meteorol.* **107**, 241-245.

Rannik, Ü., Vesala, T.: 1999, 'Autoregressive filtering versus linear detrending in estimation of fluxes by the eddy covariance method', *Bound.-Layer Meteorol.* **91**, 259-280.

Sakai, R. K., Fitzjarrald, D. R., Moore, K. E.: 2001, 'Importance of low-frequency contributions to eddy fluxes observed over rough surfaces', *J. Appl. Meteorol.* **40**, 2178-2192.

Scarborough, J. B.: 1966, *Numerical Mathematical Analysis*, 6th Edition, The Johns Hopkins Press, Baltimore.

Silverman, B. A.: 1968, 'The effect of spatial averaging on spectrum estimation' *J. Appl. Meteorol.* **7**, 168-172.

Spanier, J., Oldham, K. B.: 1987, *An Atlas of Functions*, Springer-Verlag, Berlin, 700 pp.

Verma, S. B., Ullman, F. G., Billesbach, D., Clement, R. J., Kim, J.: 1992, 'Eddy correlation measurements on methane flux in a northern peatland ecosystem' *Bound.-Layer Meteorol.* **58**, 289-304.

Vickers, D., Mahrt, L.: 1997, 'Quality control and flux sampling problems for tower and aircraft data', *J. Atmos. Oceanic Technol.* **14**, 512-526.

Villalobos, F. J.: 1996, 'Correction to eddy covariance water vapor flux using additional measurements of temperature', *Agric. For. Meteorol.* **88**, 77-83.

Webb, E. K., Pearman, G. I., Leuning, R.: 1980, 'Correction of flux measurements for density effects due to heat and water vapor transfer', *Quart. J. R. Meteorol. Soc.* **106**, 85-106.

Wilczak, J. M., Oncley, S. P., Stage, S. A.: 2001, 'Sonic anemometer tilt correction algorithms', *Bound.-Layer Meteorol.* **99**, 127-150.

Wyngaard, J. C.: 1971, 'Spatial resolution of a resistance wire temperature sensor', *Physics of Fluids* **14**, 2052-2054.

Wyngaard, J. C., Siegel, A., Wilczak, J. M.: 1994, 'On the response of a turbulent-pressure probe and the measurement of pressure transport', *Bound.-Layer Meteorol.* **69**, 379-396.

Chapter 5

LOW FREQUENCY ATMOSPHERIC TRANSPORT AND SURFACE FLUX MEASUREMENTS

Yadvinder Malhi, Keith McNaughton, Celso Von Randow
yadvinder.malhi@ouce.ox.ac.uk

Abstract

We review the issue of turbulent atmospheric transport of scalars or momentum on timescales greater than 30 minutes or 1 hour, regions of the spectrum of atmospheric motion that are not usually sampled by conventional flux measurement methodologies. We first explore what is known about the nature and timescales of turbulent transport structures in the near-surface layers of the atmosphere, and the degree to which this transport is controlled or modulated by the timescales of inner layer (shear) transport and outer boundary layer transport. We then present empirical evidence of the existence of low frequency transport by presenting data from two contrasting field studies, a shear-dominated measurement setup in Scotland, and a convection dominated measurement setup in Brazilian Amazonia. It is clear that low frequency motion can transport a significant amount of flux in measurement situations such as towers over forests in anticyclonic conditions, or in the tropics. Thereafter we explore the quantitative implications of undersampling low frequency atmospheric transport. Extending the sampling period of surface flux measurements is desirable under certain conditions, but is not without complications. We conclude by highlighting some of the dangers of extending sampling periods into the low frequency domain.

1 Introduction

The micrometeorological technique of eddy covariance aims to measure the transport of flux via turbulence between the surface and the atmosphere. In practice it samples only a part of the possible spectrum of atmospheric motion, typically time scales between one second and one hour. Transport at timescales less than one second are discussed in Chapter 4 of this book. For longer timescales an assumption is made of frequency separation — that a spectral gap exists between the operational timescales of flux transport, and longer-scale atmospheric motions. The presumed existence of such a gap is critical for measured turbulent fluxes, as it provides an upper time limit to measurements averaging and rotation periods to separate "locally meaningful" fluxes from background trends and oscillations. However, it is unclear as to whether this gap really exists. The importance of low frequency motions to energy and carbon balance studies have been highlighted in recent papers (Mahrt 1998a, 1998b, Sakai et al. 2001, Von Randow et al. 2002, Finnigan et al. 2003).

In this Chapter, we first review what is known about the nature and timescales of turbulent transport structures in the near-surface layers of the atmosphere, and the degree to which this transport is controlled by the timescales of inner layer (shear) transport and boundary layer transport. This Chapter covers some of the same territory as Mahrt (1998a), but using the perspective of the new turbulence model of McNaughton (2004b). We focus primarily here on daytime turbulent transport, the deeper complexities of nighttime transport are discussed in Chapter 8 of this book. We then present empirical evidence of the existence of low frequency transport by presenting data from two contrasting field studies, a shear dominated measurement setup in Scotland, and a convection dominated measurement setup in Brazilian Amazonia. Thereafter we explore the quantitative implications of undersampling low frequency atmospheric transport, summarizing an analysis presented by Finnigan et al. (2003). We conclude by highlighting some of the complications of extending sampling periods into the low frequency domain. The subject of low frequency transport and how to deal with it enters into the still poorly understood fields of self-organized turbulent structures and mesoscale flows, and is clearly an area of ongoing research. We begin by reviewing these issues.

2 Turbulence Structure, Eddy Sizes and Sampling Times

The question of averaging times for flux measurements clearly depends on the nature, structure and time scales of the fluctuations over which the average is taken. It is possible to approach this purely through a discussion of the statistical character of the signals to be averaged, notably through discussion of their spectra and cospectra. This section is a preliminary to that discussion, providing some insight into the turbulence processes that underlie these spectra. Since the structure of turbulence varies with height we divide the boundary layer into a series of layers, the principal ones being the "outer layer" or "mixed layer" where large convective circulations form during the day, and the "inner layer" or "surface layer" whose characteristics are dominated by shear. Within the inner layer is a roughness sublayer where the turbulence is directly affected by the surface itself. The inner and outer layers are separated by a transition layer. This has variously been called an "interaction layer" (McNaughton 2004b), a "matching layer" (Mahrt 1998a) or a "free convection layer" (e. g. Garratt 1992). These different names reflect different understandings of what this layer represents. Garratt (1992) sees it as a layer of gradual transition where proposed plume-like structures that develop progressively with height in the surface layer finally break free of shear effects while not yet being constrained by the top of the boundary layer. Mahrt (1998a) takes a more cautious approach and simply notes that layered models typically require that properties in the inner and outer layer should be mathematically matched at this height. McNaughton (2004b) describes it as a layer where two fundamentally different kinds of turbulence interact, so turbulence there is fundamentally less ordered than in the layers above and below. McNaughton's ideas are based on the few direct observations available, and we base our discussion on it.

Figure 5.1 shows four of the identifiable structural layers of a convective boundary layer. Above the boundary layer lies the free atmosphere, with an entrainment layer (another interaction layer) between this and the outer layer. These are not shown because flux measurements are not made at these levels. Figure 5.1 also shows instruments placed on towers above short and tall vegetation in relationship to these structural layers of the boundary layer. The transition layer typically begins at one or two times the Obukhov length, $|L|$, above the ground, where the Obukhov length is given by

$$L = -\frac{\Theta_v u_*^3}{kg(\overline{w'\theta_v'})_0} \tag{5.1}$$

Here g is acceleration due to gravity, $(\overline{w'\theta_v'})_0$ is the buoyancy flux at ground level, Θ_v is the virtual potential temperature in degrees Kelvin, u_* is the friction velocity, and k is the Von Karman constant. This transition layer height may vary from less than 10 m to more than 100 m depending on meteorological conditions, so instruments on a short mast over grassland will usually be in the surface layer while instruments on a tall tower over forest will be variously within the surface layer, the transition layer or even the outer layer, depending on meteorological conditions. This creates some challenging problems for flux measurement over forests.

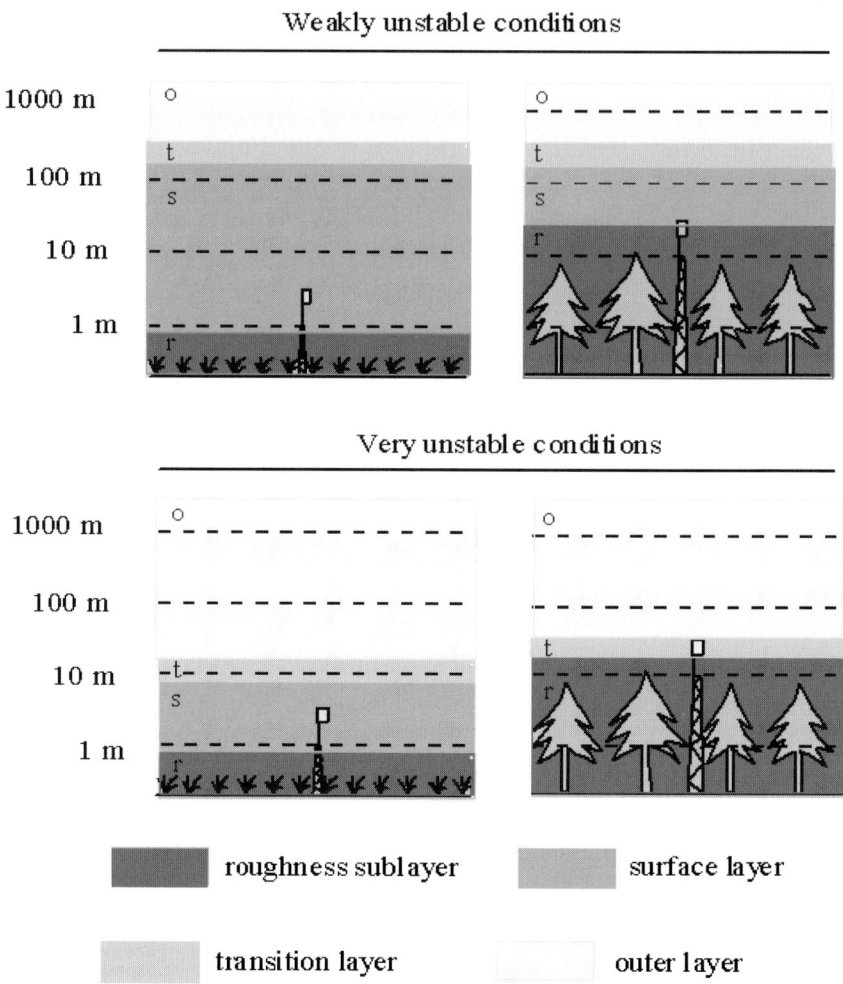

Figure 5.1. Schematic diagram of the location of flux measuring instruments, shown as rectangles on the top of each tower, in relationship to the various layers of the boundary layer in both weakly and strongly unstable conditions. Instruments mounted a few meters above low vegetation are usually within the surface layer. This layer may completely disappear over forests as the outer layer extends downwards into the layer directly influenced by the vegetation (up to at least twice forest height). Measurements from a forest tower may often be in the transition or outer layer during the day. This affects the averaging times needed to make reliable observations of scalar fluxes.

2.1 Mixed-layer (outer-layer) timescales

Over uniform land the spatial and temporal patterns of eddy motion are not imposed by the boundary conditions, upper or lower, but arise spontaneously through the self-organizing nature of the turbulent motions themselves. During the day this self-organized pattern is usually a set of more-or-less polygonal convective cells that span the whole boundary layer. These cells are about $1.5 z_i$ wide, where z_i is the height of the capping inversion. The height z_i is typically 1-3 km during the middle of the day over land, being greatest in high heat flux conditions such as continental interiors in summer. These convective cells move along with the mean wind in the outer layer. Their shapes depend on the value of the ratio of the outer and inner velocity scales, w_*/u_*, where the convective velocity scale w_* is related to the standard deviation of the wind velocity in the outer layer and is usually estimated using Deardorff's relationship

$$w_*^3 = \frac{z_i g \overline{(w'\theta_v')}_0}{\Theta_v} \tag{5.2}$$

The ratio w_*/u_* is then related to the Obukhov length through the tautological relationship

$$\frac{w_*}{u_*} = \left(-\frac{z_i}{kL}\right)^{1/3} \tag{5.3}$$

When $z_i/|L|$ is larger than about 25, the cells form a polygonal pattern with no notable alignment. At smaller values of $z_i/|L|$, when drag on the ground is more important, the convective cells become elongated in the direction of the wind and aligned, one with the next, so that they form elongated roll structures aligned with the wind. Water vapor condensing in the updrafts of these structures can form cloud streets (Etling and Brown 1993). These streets, whether made visible by clouds or not, are quasi-permanent in position and many kilometers long. Fixed sensors in the outer layer may then record steady updrafts ($\overline{w} > 0$) or steady downdrafts ($\overline{w} < 0$) over long periods, with obvious consequences for the calculation of fluxes. Very often the decrease in $z_i/|L|$ reflects an increases in $|L|$, so the surface layer grows in thickness to envelope fixed-height sensors above a forest. At $z_i/|L|$ ratios less than about 5 the boundary layer becomes near-neutral, exhibiting no large-scale convective roll structures. We might observe that the combination of deeper surface and transition layers leaves no room for outer convective structures to develop in such situations.

An important characteristic of the main convective motions in the outer layer is that they carry most of the flux, with lesser amounts carried by the smaller eddies created by internal friction and breakdown of these larger eddies. An aircraft equipped with flux sensors must fly through many tens of large eddies to sample the flux properly, so an adequate flight path would be of order 50 km long, more or less depending on prevailing conditions and the accuracy required. Flux sensors mounted on a fixed tower or suspended from a tethered balloon must sample a similar number of eddies by waiting as they blow past the instruments, so the sample period should be many tens of times z_i/U_m, where U_m is the mean wind speed in the outer part of the boundary layer. In light winds, as often encountered in the tropics or in mid-latitude anticyclones, the convective structures may pass very slowly so that a good sample is obtained only for times long compared with the evolutionary time scale of the convection cells. Values for this evolutionary time scale have not been reported, but they may depend strongly on any inhomogeneities in surface roughness, surface heat flux or topography that might cause convective cells to lock into fixed positions on the landscape. Even in favorable conditions it usually takes rather more than an hour to achieve a good average in convective conditions, and much longer if the convective cells are aligned with the wind in quasi-permanent rows or locked onto the landscape. A non-zero mean for vertical velocity may persist for hours.

Kanda et al. (2004) use Large Eddy Simulation (LES) models to calculate errors likely to be found in instrumental measurements from very tall towers over uniform ground. They do not give values of w_*/u_* for their study, but the small values of the geostrophic wind used and the cellular patterns found in their results suggests that conditions are highly convective in all cases. At low geostrophic wind speeds transport in the lower convective boundary layer self-organizes into turbulent organized structures (TOS) consisting of large areas of slow subsidence and smaller areas of updraft (Figure 5.2). As a result, a single point measurement averaged/rotated over a short time period is likely to be biased to the downdraft regions and hence underestimate the local fluxes. Simulated observations at 100 m above ground display vertical winds that do not average to zero over one-hour periods, with the net upwards or downwards flow typically accounting for half the flux when $u_* = 0.14$ m s^{-1}. This "imbalance" is reduced with (i) increasing wind speed which decreases the cross-wind diameter of the TOS and increases the sampling track in any averaging period (e. g. at $u_* = 0.3$ m s^{-1}, mean flow accounts for about 10% of vertical scalar transport on average over the simulation domain, but with values for particular points having a standard deviation of about 20% about that mean); (ii) lower measurement height increasing the influence of shear turbulence. Hence measurement points

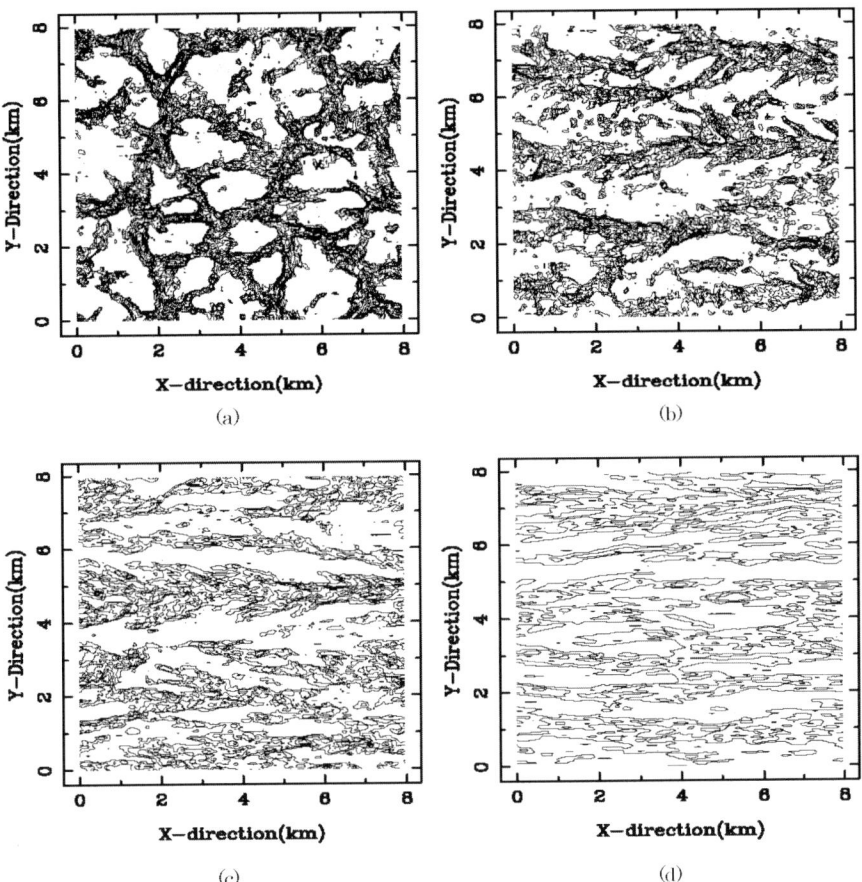

Figure 5.2. Simulations of mean vertical velocity at 100 m height in a typical mid-latitude daytime boundary layer over a flat surface, averaged over 1 hr for different geostrophic wind speeds U_g. Contours (at 0.1 m s^{-1} interval) represent positive (upwards) velocity regions; negative velocity regions are blank. (a) $U_g = 0$ m s^{-1}; (b) $U_g = 1$ m s^{-1}; (c) $U_g = 2$ m s^{-1}; (d) $U_g = 4$ m s^{-1}. Reproduced with permission of Kanda et al. (2004).

located above tall forests, and particularly in low-wind regions such as many tropical and continental interior sites, are particularly likely to be prone to such a measurement "imbalance". In these conditions flux measurements from single point measurements are inherently distorted (Figure 5.3).

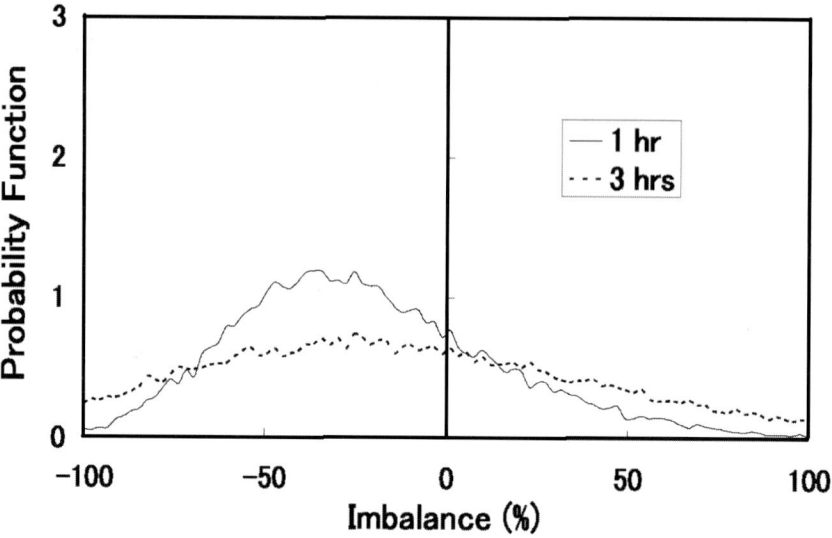

Figure 5.3. The probability distribution of the "imbalance" of fluxes measured at any point in the simulation are shown in Figure 5.2a, for geostrophic wind speed $U_g = 0$ m s^{-1}, height $z = 100$ m. The two curves represent averaging periods of 1 and 3 hr. Measurements at any point are biased to underestimate surface fluxes. Increasing the average period reduces the systematic error in the "imbalance", but broadens the distribution and hence the random uncertainty in any single measurement. Reproduced with permission of Kanda et al. (2004).

2.2 Surface-layer (inner-layer) timescales

Near the ground the flux-carrying eddies have a rather different character, though what that character is has become the subject of debate. Here we base our discussion on the new model of McNaughton (2004b) in which the turbulence consists of large-scale wedge-like structures that are aligned with the wind, along with the detached breakdown products of these. The turbulence is again self-organizing, just like the cellular structures in the outer layer, but here the preferred structures are upscale cascades of TEAL (Theodorsen ejection amplifier-like) structures, which "compete" for space so that only the best formed and most powerful continue on at each scale. This kind of turbulence is driven by the shear, so its velocity scale is u_*. Though the value of u_* changes with stability it seems that the structure of the shear turbulence does not. This property is again like that of the turbulent outer layer where the velocity scale of the convective cells varies with stability while the polygonal structure and the length scale of turbulence does not (at least while w_*/u_* and z_i are held constant). This model is consistent with — indeed it is based on — the spectral observations from Kansas (Kaimal et al. 1972). It contradicts the basic hypothesis of Monin-Obukhov similarity on which much of our theory of the surface layer is built. Time will judge whether the new model is successful, but for the moment it simplifies our discussion by removing the need to consider the effects of stability acting within the surface layer, at least when we know u_* directly. Turbulence in the surface layer scales simply on z and u_*.

To get a good sample of the dominant flux-carrying eddies, with widths about $2z$, we must allow a large number of them to pass our fixed measuring position. A simple estimate is that this should be a hundred times $z/\overline{u}$, which is a very few minutes in most cases. Experience tells us that this is a considerable underestimate and there are several reasons for this. One is that the flux-carrying coherent structures are not randomly dispersed, so our sample must take account of the scale of the aggregates (i. e. whole wedge structures) which are about ten times longer than our estimate of about $2z$ (McNaughton 2004a). Another reason is that flux transport is affected by eddies of a greater range of sizes here than in the outer layer. In the outer layer there are no eddies much wider than $1.5z_i$, while in the surface there are eddies of all sizes, right up to those as tall as the surface layer itself, and all of these transport at least some momentum and scalars. Our averaging period must be long enough to sample not just the dominant eddies but all the significant flux-carrying eddies. As suitable averaging period must be at least several tens of minutes long, perhaps even an hour long for good results.

Another factor is the modulating effects of the outer convection on the turbulence in the surface layer. The outer-layer convection constitutes a variable driving of the whole surface layer. A large-scale gust from the outer layer is equally a large cohort of TEAL structures moving along with enhanced speed and transmitting an enhanced momentum flux towards the ground. The direction, speed and power of the TEAL cascades within the surface layer will therefore all vary with the wind at the top of the surface layer, as will the momentum flux to the ground. Observations of momentum flux from a fixed tower will not represent area means unless sampled on an *outer time scale*, not the inner one. We expect poor agreement in hourly observations of u_* from towers spaced a few hundred meters apart in homogeneous terrain. The situation is easier for the scalar fluxes, most of which are more strongly controlled at source rather than by the varying wind overhead. For example, photosynthesis is controlled by radiation receipt and biochemical and physiological processes within the canopy, both of which are insensitive to wind speed. The flux of water vapor is somewhat sensitive to wind speed, depending on the value of the decoupling parameter Ω calculated with values at reference height z_s (Jarvis and McNaughton 1986), but only for wet canopies is it important. Averaging times for source-limited scalar fluxes depend little on outer time scales. This difference accounts for the reported different averaging times required for momentum and scalars in the Kansas experiment (Wyngaard 1973). Wyngaard shows the variance of heat flux measurements from one-hour runs to be about 8%, with only small dependence on instability ($-z/L$), while the sensitivity of momentum flux ranged from 10% to about 80% with a strong dependence on instability. Wyngaard used $-z/L$ to measure instability while our discussion suggests that $-z_i/L$ should be used.

The layered structure discussed above is not relevant when the outer turbulence dies away at night. In the first part of the night we have, typically, a fully-turbulent state where the turbulence is similar to that in the daytime surface layer, but buoyancy now opposes these motions and saps their energy. Momentum transfer then decreases progressively. This kind of structure collapses altogether unless the flow is maintained by strong pressure gradients or katabatic forcing. If not then a variety of other phenomena may appear, some creating intermittency in the turbulence with time scales up to an hour or more. These processes are not well described and are discussed in Chapters 8 and 9 of this book. For the weakly stable case flux sampling times can be a little shorter than in the daytime surface layer.

The roughness sublayer is a subdivision of the surface layer and many of the above comments apply equally to it. Near canopy top the turbulence has many of the characteristics of a mixing layer (Raupach et al. 1996), and the main eddies scale on the canopy dimensions $(h-d)$, where h is tree height and d is the displacement height of the forest wind profile. This structure is not well known, but it is known that spectra and cospectra have more peaked shapes than the standard Kansas forms. Following the methods of McNaughton (2004a) we can therefore surmise that tendency of these eddies to align into long wedge-like structures is less pronounced than in the surface layer. Good time samples should be easier to obtain in the surface layer, but the length scale used to calculate these is $(h-d)$ rather than $(z-d)$.

A theme running through the above comments is that eddy flux calculations made on short data runs may not fairly represent true long-term averages. This is so whether the flux itself varies on a rather long time scale, or whether the statistical sampling of the flux-carrying eddies is insufficient. In the former case the actual local flux is a poor sample of the required area average flux; this cannot be remedied by any analysis based solely on the data from that interval. In the latter case the local flux during the measurement interval is not in question, but some of the larger eddies carrying it have been poorly sampled.

3 Empirical Evidence of Low Frequency Flux Transport

3.1 Wavelet spectral analyses of turbulent fluxes in Scotland and Amazonia

To demonstrate the variable contribution of processes occurring on different scales to the surface layer fluxes, in this study we apply a Haar wavelet transform on the turbulent signals measured at two different sites: a rain forest in south west Amazonia (Rebio Jaru; Von Randow et al. 2002) and a sitka spruce coniferous forest in Scotland (Griffin; Chapter 4). The wavelet transform (WT) is a powerful mathematical analysis tool, which permits an evolutionary spectral study of turbulent atmospheric signals (Daubechies 1998; Farge 1992). The wavelet analyses were done following a similar methodology as Katul and Parlange (1994) and Von Randow et al. (2002).

After application of the Haar wavelet to the data from the two sites, the scale covariances of vertical wind velocity and scalars were calculated and the partial contribution of each scale to the total covariance was determined at each record. The results are presented in Figure 5.4 (daytime) and Figure 5.5 (nighttime). The x-axes represent the spatial scales, which are estimated using the average wind velocities and the assumption of Taylor's hypothesis, similar to the method applied by Von Randow et al. (2002). Above the x-axes, approximate time labels are also included to illustrate the time scales of the processes.

Comparing the average daytime results from Jaru and Griffin (Figure 5.4), it is apparent that the contribution from the largest scales (lower frequencies) are more important at Jaru than at Griffin. At Rebio Jaru, the peak of contribution to the covariances happens on scales of 100-500 m (that correspond approximately to scales of 1 to 5 min), while at Griffin scales less that 100 m (usually corresponding to scales of less than 1 min) dominate the transport of carbon dioxide. One other noticeable difference between the two sites is that at Jaru the variation from the average scale dependence is higher, especially at longer scales (low frequencies). On scales longer than 1 km, the low frequency motions influences can be of either sign (Figure 5.4 top panel), clearly not related to processes of the surface layer only. These low frequency processes can include deep convection, large roll vortices and local circu-

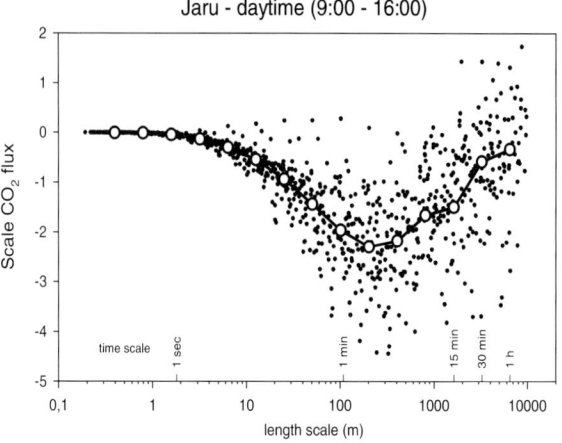

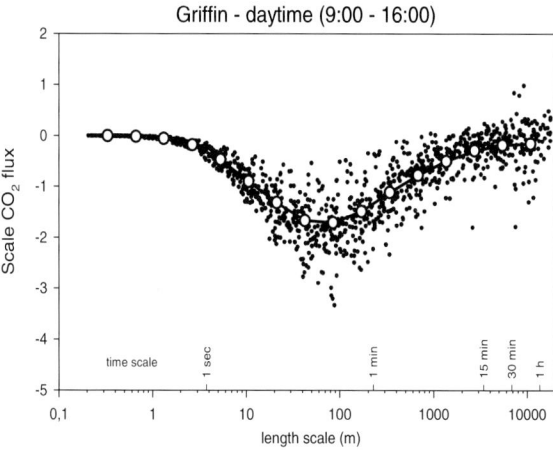

Figure 5.4. Haar wavelet cospectra of daytime CO_2 fluxes at a tropical site (top panel; Jaru, Brazil) and (b) a maritime mid-latitude site (bottom panel; Griffin, Scotland). The solid line represents binned averages; length and time scales are indicated on the x-axis.

lations induced by topography or surface heterogeneity. At the Brazilian site, a modest but significant amount of turbulent transport occurs at time scales beyond 30 minutes, or even beyond one hour.

During stable conditions (nighttime) there is no clear timescale for flux transport at the Brazilian site (Figure 5.5 top panel), and a great

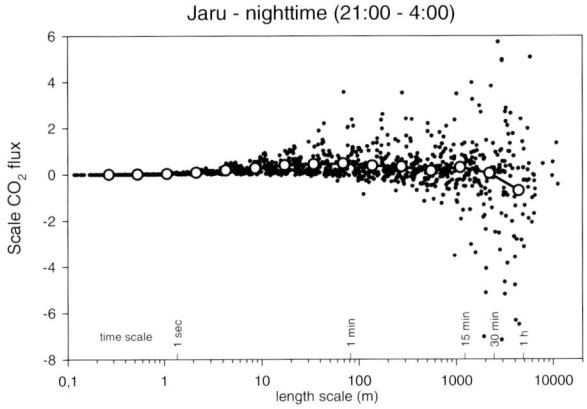

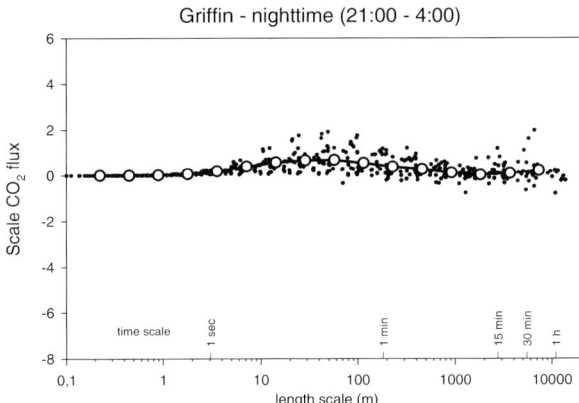

Figure 5.5. Haar wavelet cospectra of nighttime CO_2 fluxes at a tropical site (top panel; Jaru, Brazil) and a maritime mid-latitude site (bottom panel; Griffin, Scotland). The solid line represents binned averages; length and time scales are indicated on the x-axis.

degree of variance in the spectra. Boundary-layer and mesoscale flows clearly dominate the nighttime turbulent flux. In contrast, the spectra are fairly consistent at the relatively windy Scottish site (Figure 5.5 bottom panel).

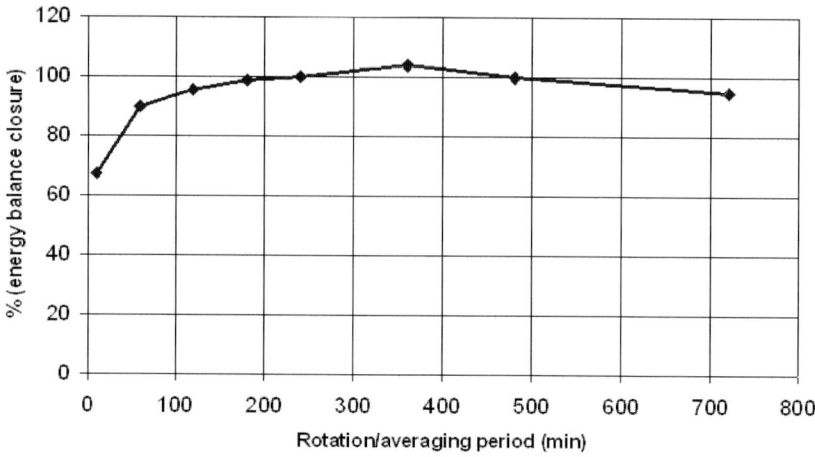

Figure 5.6. The relationship between long-term energy balance closure and the averaging/rotation period of the flux calculations, for a site over a tropical rainforest near Manaus, Brazil. Derived from Malhi et al. (2002).

3.2 Low frequency transport and energy balance

The lack of energy balance closure in eddy covariance studies is a widespread feature of flux measurements over forests (Wilson et al. 2002), suggesting the presence of a widespread problem, whether instrumental or methodological. One possible explanation is that latent and sensible heat flux is being missed by the flux measurements, through either inadequate spatial sampling or inadequate sampling of the frequency domain.

Malhi et al. (2002) explored this phenomenon for flux measurements above a forest near Manaus in central Amazonia (Figure 5.6). They found that extending the rotation/averaging period of the measurements from 1 hr up to 4 hr improved the energy balance closure to about 100%; increasing the rotation/averaging period further resulting in no further increase in mean fluxes, although variance increased substantially. This result suggests that, for some sites at least, low frequency transport may "solve" the energy balance problem, for ensemble-averaged data at least. The general applicability of this approach remains uncertain, however, as Kruijt (pers comm.) and Malhi and Iwata (unpublished data) find that the low frequency component at other sites is not sufficient to account for all the "missing" flux.

To summarize this section, there is evidence of significant flux transport at low frequencies at some measurement sites; in particular sites in the tropics with tall measurement towers, low mean wind speeds and deep convective boundary layers. In the next section we examine how undersampling these low frequency fluxes using standard eddy covariance analysis can affect measured fluxes.

4 Effect of Standard Flux Calculations on Low Frequency Flux Terms

In this section we explore the effect that the coordinate rotation used in eddy covariance analyses has on calculated fluxes. Finnigan et al. (2003) recently presented a revision of the theory of measurement of turbulent fluxes in terms of mass balance equations. First they considered an idealized case, where the long-term ensemble-averaged flow is horizontally homogeneous, the mean wind vector is always confined to the $x - z$ plane, so that $\overline{v} = 0$ and only the inclination of the velocity vector, $\alpha = \tan^{-1}(\overline{w}/\overline{u})$ changes from period to period. To transform the long-term vertical covariance $\overline{wc}_{LT}$ in any period to short-term 'rotated every period' coordinates in which $\overline{w} = 0$, coordinates are rotated through an angle α given by,

$$\alpha = \tan^{-1} \frac{\overline{w}'}{<\overline{u}> + \overline{u}'} \qquad (5.4)$$

where $<>$ is the ensemble mean over all periods, $\overline{w}' = \overline{w} - <\overline{w}>$, and $\overline{u}' = \overline{u} - <\overline{u}>$.

Hence, the horizontal and vertical components of the long-term, ensemble-averaged flux vector, denoted by subscript LT, can be expressed in terms of components, denoted by subscript, R, in the "rotated-every-period" coordinate frame as (Finnigan et al. 2003, equation 26)

$$\begin{aligned}(\overline{uc})_R &= (\overline{uc})_{LT} \cos\alpha - (\overline{wc})_{LT} \sin\alpha \\ (\overline{wc})_R &= (\overline{wc})_{LT} \cos\alpha + (\overline{uc})_{LT} \sin\alpha \end{aligned} \qquad (5.5)$$

Even in horizontally homogeneous terrain, the flow field is only horizontally homogeneous when averaged over a period much longer than that of any significant temporal perturbation to the flow. When the duration of an individual averaging period, T, is comparable to that of significant atmospheric motions, then in any one period, horizontal flux divergence may be important. Many periods must be averaged to ensure the canceling out of transient vertical advection events that in reality contribute nothing to the vertical flux because they merely balance simultaneous but unmeasured transient horizontal advection events. To ensure this averaging is done rigorously, mass balances over any time period must be calculated in a single coordinate system that applies to the entire period.

What is the precise effect of rotating the vector basis so that $\overline{w}$ is forced to zero in each period? In particular, what is the effect on the measured covariance in any one averaging period of rotating coordinates so that $\overline{v} = \overline{w} = 0$? *Dividing the signals into a set of periods of length T, block averaging in each period and then subtracting the block averaged values $\overline{w}$ and $\overline{c}$ is equivalent to high-pass filtering the signal with a boxcar filter function of width T.* This is precisely what is achieved when we rotate coordinates each period so that $\overline{w}$ is forced to zero. In effect we have thrown away the part of the covariance carried by motions of frequency lower than the (rather leaky) cut-off frequency of the boxcar filter function.

However, in addition, the coordinate rotation itself distorts the covariance carried by frequencies higher than the boxcar cut-off by folding into $(\overline{wc})_R$ some of the streamwise and lateral fluxes $(\overline{wc})_{LT}$ and $(\overline{vc})_{LT}$, as we have shown in Equation 5.5 for the case where the low frequency flow is confined to the $x - z$ plane. A simple algebraic example of the distortion to be expected is presented in Appendix 2 of Finnigan et al. (2003).

In summary, rotating coordinates every period effectively high-pass filters the signal so that contributions to the covariance from atmospheric motions of period longer than T are lost but the rotation itself folds some of the $(\overline{wc})_{LT}$ and $(\overline{vc})_{LT}$ flux into $(\overline{w'c'})_R$ in an essentially unpredictable way.

5 Complications

The energy balance closure in Figure 5.6 suggests that in some sites consideration of low frequency transport can improve the flux measurements to the point of full energy balance closure. If this were a general principle that could be applied to all sites, then it would appear that the problem of poor energy balance closure at eddy covariance sites would have been "solved". However, there are good grounds for skepticism about the wider applicability of this result.

The problem revolves around whether these low frequency fluxes are "locally meaningful" or represent features of the wider landscape that are not related to the local surface. Kanda et al. (2004) demonstrated in their LES model simulation that, although the systematic bias decreased when turbulent fluxes are averaged over longer time periods, the variance increases greatly (Figure 5.3). Hence any single measurement period is vulnerable to random advective fluxes, and becomes increasingly difficult to interpret in terms of local surface fluxes.

Another problem arises in complex terrain, or in the presence of fixed pressure fields caused by local inhomogeneities (e. g. land-water interfaces, or crop-forest mosaics). In this situation, a variation in wind direction can result in a low frequency covariance that has nothing to do with flux transport (Finnigan et al. 2003, Chapter 2). Interestingly, Kanda et al. (2004) showed that moderate inhomogeneity (ca 5%) can actually reduce bias by dampening the self-organization of TOS, but greater degrees of inhomogeneity generate local circulations and enhance the bias.

6 Conclusions

It is important to consider transport at time scales up to at least $10z_i/U_m$, where U_m is the mean wind speed in the outer part of the boundary layer. This corresponds to length scales of about 10-50 km. But in complex topography or variable landscapes, it is tricky to disentangle what we want (low frequency CBL turbulent transport) from what we do not want (wind direction covariance). There is clearly transport of flux at these low frequencies which can explain the failure of eddy covariance systems to fully capture fluxes. However, separating the locally meaningful fluxes from wider-scale advection may prove a challenge.

7 References

Daubechies, I.: 1998, 'Recent results in wavelet applications', *J. Electronic Imaging*, **7**, 719-724.

Etling, D., Brown, R. A.: 1993, 'Roll vortices in the planetary boundary layer: a review', *Bound.-Layer Meteorol.*, **65**, 215-248.

Farge, M.: 1992, 'Wavelet transforms and their applications to turbulence', *Annual Rev. Fluid Mech.*, **24**, 395-457.

Finnigan, J., Clement, R., Malhi, Y., Leuning, R., Cleugh, H. A.: 2003, 'A re-evaluation of long-term flux measurement techniques. Part I: Averaging and coordinate rotation', *Bound.-Layer Meteorol.*, **107**, 1-48, 2003.

Garratt, J. R.: 1992, *The Atmospheric Boundary Layer*. Cambridge University Press, Cambridge, 316 pp.

Jarvis, P. G., McNaughton, K. G.: 1986, 'Stomatal control of transpiration: scaling up from leaf to region', *Adv. Ecol. Res.*, **15**, 1-49.

Kaimal, J. C., Wyngaard, J. C., Izumi, Y., Coté, O. R.: 1972, 'Spectral characteristics of surface-layer turbulence', *Quart. J. R. Meteorol. Soc.*, **98**, 563-589.

Kanda, M. Inagaki, A., Letzel, M. O., Raasch, S., Watanabe, T.: 2004, 'LES study of the energy imbalance problem with eddy covariance fluxes', *Bound.-Layer Meteorol.*, **110**, 381-404.

Katul, G. G., Parlange, M., B.: 1994, 'On the active role of temperature in surface-layer turbulence', *J. Atmos. Sci.*, **51**, 2181-2195.

Mahrt, L.: 1998a, 'Surface fluxes and boundary layer structure', In: (Eds.) Holtslag, AAM, Duynkerke, PG, *Clear and Cloudy Boundary Layers*, Royal Netherlands Academy of Arts and Sciences, Amsterdam, 113-128.

Mahrt, L.: 1998b, 'Flux sampling errors for aircraft and towers', *J. Atmos. Oceanic Tech.*, **15**, 416-429.

Malhi, Y., Pegoraro, E., Nobre, A. D., Pereira, M. G. P., Grace, J., Culf, A. D., Clement, R.: 2002, 'Energy and water dynamics of a central Amazonian rain forest', *J. Geophys. Res.*, **107**, Art. No. 8061.

McNaughton, K. G.: 2004a, 'Attached eddies and production spectra in the atmospheric logarithmic layer', *Bound.-Layer Meteorol.*, **111**, 1-18.

McNaughton, K. G.: 2004b, 'Turbulence structure of the unstable atmospheric surface layer and transition to the outer layer', *Bound.-Layer Meteorol.*, in press.

Raupach, M. R., Finnigan, J. J., and Brunet, Y.: 1996, 'Coherent eddies and turbulence in vegetation: the mixing layer analogy', *Bound.-Layer Meteorol.*, **78**, 351-382.

Sakai, R. K., Fitzjarrald, D. R., Moore, K. E.: 2001, 'Importance of low-frequency contributions to eddy fluxes observed over rough surfaces', *J. Appl. Meteorol.*, **40**, 2178-2192.

Von Randow, C., Sá, L. D. A., Gannabathula, P. S. S. D., Manzi, A. O., Arlino, P. R. A., Kruijt, B.: 2002, 'Scale variability of atmospheric surface layer fluxes of energy and carbon over a tropical rain forest in southwest Amazonia. I. Diurnal conditions', *J. Geophys. Res.*, **107**, 8062, doi:10.1029/2001JD000379.

Wilson, K., Goldstein, A., Falge, E., Aubinet, M., Baldocchi, D., Berbigier, P., Bernhofer, C., Ceulemans, R., Dolman, H., Field, C., Grelle, A., Ibrom, A., Law, B. E., Kowalski, A., Meyers, T., Moncrieff, J., Monson, R., Oechel, W., Tenhunen, J., Valentini, R., Verma, S.: 2002, 'Energy balance closure at FLUXNET sites', *Agric. For. Meteorol.*, **113**, 223-243.

Wyngaard, J. C.: 1973, 'On surface-layer turbulence', In: (Ed.) Haugen, DA, *Workshop on Micrometeorology*, American Meteorological Society, 101-149.

Chapter 6

MEASUREMENTS OF TRACE GAS FLUXES IN THE ATMOSPHERE USING EDDY COVARIANCE: WPL CORRECTIONS REVISITED

Ray Leuning
Ray.Leuning@csiro.au

Abstract This Chapter re-examines theory developed by Webb, Pearman and Leuning (1980, Quarterly Journal of the Royal Meteorological Society, 106, 85-100) to calculate fluxes of trace gas constituents in the atmosphere using the eddy covariance technique. The original theory for one-dimensional flow over homogeneous terrain is extended to three-dimensional flow over inhomogeneous terrain. The equations are relatively simple when concentrations are expressed as mixing ratios per unit of dry air. Advective mass fluxes are written as products of fluxes of dry air and gradients in mixing ratio, while turbulent eddy fluxes requires the covariance of wind speeds and mixing ratios. Theory developed by WPL for one dimensional flows is applicable for the vertical eddy flux.

1 Introduction

The eddy covariance technique is used widely to measure the net exchanges of heat, mass and momentum between the earth's surface and the atmosphere (Baldocchi et al. 2001). Before publication of the paper by Webb et al. (1980) (WPL hereafter), the vertical turbulent flux density of a constituent c was calculated as $\overline{F}_c = \overline{w'c'_c}$, the covariance between fluctuations in the vertical velocity, w' and the density c'_c. WPL showed that this gave incorrect estimates of $\overline{F}_c$ because fluctuations in c_c can result from fluctuations in water vapor density and temperature which are not associated with the net transport of c. These errors are particularly severe for trace constituents such as CO_2. The original WPL theory strictly only applies to steady, one-dimensional flow over horizontally homogeneous terrain and hence may not be suitable for the more typical flux measurement installation in inhomogeneous terrain. Further theoretical work is thus warranted.

Recent papers by Kramm et al. (1995), Sun et al. (1995), Paw U et al. (2000), Massman and Lee (2002) and Fuehrer and Friehe (2002) have re-examined the conservation equations used to calculate net exchanges of mass and energy between the earth's surface and the atmosphere for surface boundary layer flows in inhomogeneous terrain. In doing so they revised the theory developed by WPL and introduced extra terms into the equations. This Chapter also examines the theory used to calculate fluxes using the eddy covariance technique and shows that the original WPL theory is still applicable for the vertical component of the eddy fluxes and that the resulting equations are particularly simple when concentrations are expressed as mixing ratios per unit of dry air.

Section 2 develops the conservation equations for the various constituents of moist air to generalize the one-dimensional conservation equation used by WPL; Section 3 utilizes the results for the special case of steady, one-dimensional, horizontally homogeneous flow to derive a key result of WPL; Section 4 considers the general case of non-steady flows in non-homogeneous terrain and discusses the components of the mass balance equation; Section 4 also discusses the case of steady, horizontally homogeneous, on-dimensional flows; Section 5 discusses practical aspects of calculating flux densities using closed- and open-path gas analyzers; and Section 6 draws some conclusions.

2 Conservation Equations for Moist Air and Trace Constituents

Consider a fixed control volume dV containing moist air with molar concentration $c = c_d + c_v + c_c$ (mol m^{-3}), in which c_d, c_v and c_c are the molar concentrations of dry air, water vapor and a trace constituent, c. (Note that while molar quantities are used in this Chapter, all equations can be written in mass units, making suitable allowances for the molecular mass of the various constituents when applying the gas laws.)

The molar conservation equation for all gas components in dV is

$$\frac{\partial c}{\partial t} + \nabla . \mathbf{F} = S_v + S_c \qquad (6.1)$$

where $\partial c / \partial t$ is the rate of change of molar concentration of air in dV, $\mathbf{F}$ is the total flux density vector on the surfaces of the control volume and S_v, S_c (mol m^{-3}s^{-1}) are the source/sinks for water vapor and trace constituent within dV. We assume that there is no source or sink of dry air within dV. Equation 6.1 may be written as

$$\frac{\partial c}{\partial t} + \nabla.\mathbf{u}c = S_v + S_c \quad (6.2)$$

where the velocity vector $\mathbf{u}$ has components $\{u, v, w\}$ in the orthogonal directions $\{x, y, z\}$. The velocity vector is *defined* as $\mathbf{u} = \mathbf{F}/c$, i. e. the molar flux density vector for moist air divided by the total molar concentration of moist air. Fluxes in the atmosphere due to molecular diffusion are assumed to be negligible.

The conservation equation for the constituent c is

$$\frac{\partial c_c}{\partial t} + \nabla.\mathbf{u}c_c = S_c \quad (6.3)$$

Equation 6.3 may also be written as

$$\frac{\partial c_d \chi_c}{\partial t} + \nabla.(\mathbf{u}c_d \chi_c) = S_c \quad (6.4)$$

where χ_c is the mixing ratio of c relative to dry air $\chi_c = c_c/c_d$.

We next use Reynolds decomposition to separate quantities into mean and fluctuating components, and then take the time-average, represented by the overbar, to give

$$\frac{\overline{\partial(\bar{c}_d + c'_d)(\bar{\chi}_c + \chi'_c)}}{\partial t} + \nabla.[\overline{(\bar{\mathbf{u}} + \mathbf{u}')(\bar{c}_d + c'_d)(\bar{\chi}_c + \chi'_c)}] = \bar{S}_c \quad (6.5)$$

Expanding the terms in this equation yields

$$\bar{c}_d \frac{\overline{\partial \chi}}{\partial t} + \bar{\chi}_c \frac{\overline{\partial c_d}}{\partial t} + \nabla.[\bar{\chi}_c(\bar{\mathbf{u}}\bar{c}_d + \overline{\mathbf{u}'c'_d}) + \bar{c}_d\overline{\mathbf{u}'\chi'_c} + \bar{\mathbf{u}}\,\overline{c'_d\chi'_c} + \overline{\mathbf{u}'c'_d\chi'_c}] = \bar{S}_c \quad (6.6)$$

where terms such as $\overline{\bar{w}\,\bar{c}_d\,\chi'} = 0$ by definition.

To proceed, we need to show that the last two terms on the left of Equation 6.6 are small compared to the others. The covariance between c_d and χ_c will be zero when fluctuations in c_d result from fluctuations in temperature and pressure, since these do not alter the mixing ratios of the constituents(i. e. $\overline{\chi'_c P'} = \overline{\chi'_c T'} = 0$ and $\overline{\chi'_c c'_v}$ is small). Fluctuations in moisture content will change both c_d and χ_c but these are expected to have only a small influence on the covariance $\overline{c'_d \chi'_c}$. For similar reasons, the triple moment $\overline{\mathbf{u}'c'_d \chi'_c}$ will also be very small (WPL). With these assumptions and noting that $\overline{\mathbf{u}c_d} = \bar{\mathbf{u}}\,\bar{c}_d + \overline{\mathbf{u}'c'_d}$, Equation 6.6 becomes

$$\bar{c}_d \frac{\overline{\partial \chi_c}}{\partial t} + \bar{\chi}_c [\frac{\overline{\partial c_d}}{\partial t} + \nabla.\overline{\mathbf{u}c_d}] + \overline{\mathbf{u}c_d}.\nabla \bar{\chi}_c + \nabla.(\bar{c}_d\overline{\mathbf{u}'\chi'_c}) = \bar{S}_c \quad (6.7)$$

Finally, there are no sources or sinks of dry air in the control volume and thus

$$\frac{\overline{\partial c_d}}{\partial t} + \nabla . \overline{\mathbf{u} c_d} = 0 \tag{6.8}$$

Substitution of this expression into Equation 6.7 yields

$$\bar{c}_d \frac{\overline{\partial \chi_c}}{\partial t} + \overline{\mathbf{u} c_d} . \nabla \overline{\chi}_c + \nabla . (\overline{c_d \mathbf{u}' \chi'_c}) = \overline{S}_c \tag{6.9}$$

Equation 6.9 states that the source/sink for constituent c equals the sum of: 1) the time rate of change of the mixing ratio χ_c in dry air, 2) the dot product of the mean flux of dry air and the gradient of χ_c at each surface of the volume, and 3) the divergence of the turbulent flux of mixing ratio multiplied by the mean density of dry air. Equation 6.9 is the non-steady, three dimensional version of an expression for the eddy flux derived by WPL for steady, one-dimensional, horizontally homogeneous flows. Equation 6.9 can be shown to be a condensed version of Equation B22 in Massman and Lee (2002) provided we assume that $\nabla . \mathbf{u} = 0$. Paw U et al. (2000) also derived a similar form of the conservation equation. Note that contrary to Equation B22 of Massman and Lee (2002) the time-averaging operator applies the time derivative $\partial \chi_c / \partial t$, not just to χ_c.

3 Non-steady, Three Dimensional Flow

Equations 6.1 and 6.9 strictly refer to an infinitesimal control volume dV while in practice we wish to measure the net exchanges of heat, water vapor and trace constituents between the earth's surface and the atmosphere. As detailed in Finnigan et al. (2003), we need to write the conservation equations for a finite control volume representative of a surface patch of area A and height h of the measuring instruments (Figure 6.1).

Integrating Equation 6.9 horizontally over A and vertically over h we obtain

$$\int_0^h \int_{x-L}^{x+L} \int_{y-L}^{y+L} \bar{c}_d \frac{\overline{\partial \chi_c}}{\partial t} dx dy dz$$

$$+ \int_0^h \int_{x-L}^{x+L} \int_{y-L}^{y+L} \left[\overline{uc_d} \frac{\partial \overline{\chi}_c}{\partial x} + \overline{vc_d} \frac{\partial \overline{\chi}_c}{\partial y} + \overline{wc_d} \frac{\partial \overline{\chi}_c}{\partial z} \right] dx dy dz$$

$$+ \int_0^h \int_{x-L}^{x+L} \int_{y-L}^{y+L} \left[\frac{\partial \overline{c_d u' \chi'_c}}{\partial x} + \frac{\partial \overline{c_d v' \chi'_c}}{\partial y} + \frac{\partial \overline{c_d w' \chi'_c}}{\partial z} \right] dx dy dz$$

$$= <\overline{S}_c> \tag{6.10}$$

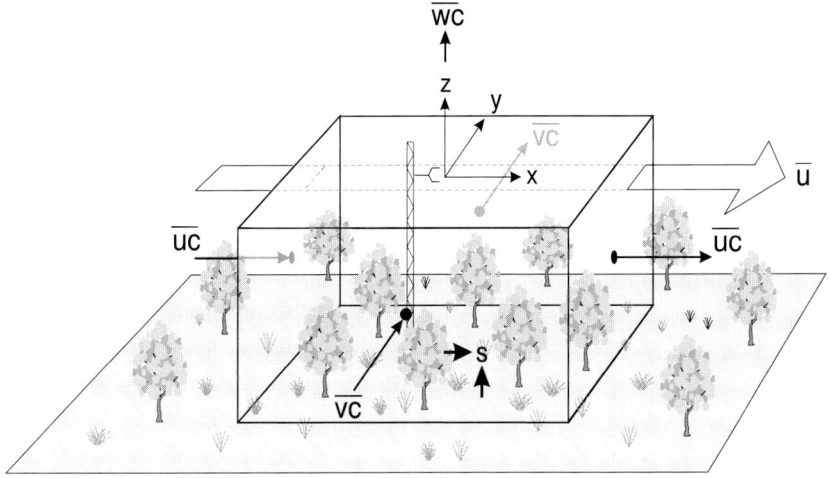

Figure 6.1. Cartesian control volume placed over a vegetated surface.

In writing this equation we have assumed a rectangular Cartesian coordinate frame with the lower boundary of the control volume placed on the ground. The right hand term represents the volume-integral of the source of c between the ground surface and the atmosphere at height h,

$$<\overline{S}_c> = \int_0^h \int_{x-L}^{x+L} \int_{y-L}^{y+L} \overline{S}_c \mathrm{d}x\mathrm{d}y\mathrm{d}z$$

When measurements are made on a single tower we are unable to measure the spatial averages that appear in Equation 6.10 and we are then obliged to add extra information. The first step usually adopted is to define a coordinate system in which $\overline{v} = \overline{w} = 0$ (strictly $\overline{vc_d} = \overline{wc_d} = 0$) and where the x-axis is aligned with the mean wind for each averaging period (e. g. McMillen 1988). Methods to define consistent, long-term coordinates have been described by Paw U et al. (2000) and Wilczak et al. (2001) and further discussed by Finnigan (2004) and in Chapter 3. For present purposes, we assume that a suitable coordinate framework has been defined and that it is possible for mean fluxes of dry air to be non-zero through any of the surfaces of the control volume, except at the ground ($\overline{wc_d}|_0 = 0$). Of course, it is also necessary to satisfy Equation 6.8 as applied to the finite control volume of Figure 6.1. The coordinate system in the subsequent analysis has been aligned with the mean wind direction so that $\overline{vc_d} = 0$.

When the divergences of the horizontal eddy fluxes are small compared to the vertical,

$$\frac{\partial \overline{c_d u' \chi'_c}}{\partial x}, \frac{\partial \overline{c_d v' \chi'_c}}{\partial y} \ll \frac{\partial \overline{c_d w' \chi'_c}}{\partial z} \quad (6.11)$$

then Equation 6.10 becomes

$$\int_0^h \int_{x-L}^{x+L} \int_{y-L}^{y+L} \overline{c_d} \frac{\partial \overline{\chi_c}}{\partial t} dx dy dz$$

$$+ \int_0^h \int_{x-L}^{x+L} \int_{y-L}^{y+L} \overline{uc_d} \frac{\partial \overline{\chi_c}}{\partial x} dx dy dz$$

$$\int_{x-L}^{x+L} \int_{y-L}^{y+L} \int_0^h \left[\overline{wc_d} \frac{\partial \overline{\chi_c}}{\partial z} + \frac{\partial \overline{c_d w' \chi'_c}}{\partial z} \right] dz dx dy$$

$$= <\overline{S_c}> \quad (6.12)$$

Equation 6.12 may be approximated by

$$\int_0^h \int_{x-L}^{x+L} \int_{y-L}^{y+L} \overline{c_d} \frac{\partial \overline{\chi_c}}{\partial t} dx dy dz$$

$$+ \int_0^h \int_{y-L}^{y+L} \overline{uc_d} (\overline{\chi_c}|_{+L} - \overline{\chi_c}|_{-L}) dy dz$$

$$\int_{x-L}^{x+L} \int_{y-L}^{y+L} \left[\overline{wc_d}|_h (\overline{\chi_c}|_h - <\overline{\chi_c}>) + \overline{c_d w' \chi'_c}|_h \right] dx dy$$

$$= <\overline{S_c}> \quad (6.13)$$

in which $<\overline{\chi_c}> h = \int_0^h \overline{\chi_c} dz$. The vertical advection term was approximated using the product rule of integration and the assumption that $\partial \overline{wc_d}/\partial z \simeq \overline{wc_d}|_h/h$ (Lee 1998, Finnigan 1999). This approximation is unnecessary if the variation of $\overline{wc_d}$ and $\partial \overline{\chi_c}/\partial z$ with height are known.

The mean horizontal mass flux of dry air $\overline{uc_d}$ in Equation 6.13 is not normally measured, but as demonstrated below, it is closely approximated by $\overline{u}\,\overline{c_d}$. The mean streamwise velocity is defined as $\overline{u} = \overline{F}_{t,x}/\overline{c} = (\overline{F}_{d,x} + \overline{F}_{v,x} + \overline{F}_{c,x})/(\overline{c_d} + \overline{c_v} + \overline{c_c})$, where $\overline{F}_{t,x}$ is the total flux of air in the x direction, and $\overline{c}$ is the total mean concentration. The mean horizontal flux of dry air is $\overline{F}_{d,x} = \overline{uc_d}$. Combining these definitions gives

$$\frac{\overline{c_d}\,\overline{u}}{\overline{c_d u}} = \frac{\overline{c_d}(\overline{F}_{d,x} + \overline{F}_{v,x})}{\overline{c}\,\overline{F}_{d,x}} = \frac{1 + \overline{F}_{v,x}/\overline{F}_{d,x}}{1 + \overline{\chi}_v} \simeq \frac{1 + \overline{\chi}_v}{1 + \overline{\chi}_v} = 1 \quad (6.14)$$

This derivation assumes that the horizontal flux of the trace constituent c, is small compared to dry air and water vapor, and that the ratio of the advective flux of water vapor to that of dry air is equal to the mixing ratio for water vapor. Thus to a close approximation

$$\overline{c_d u} = \overline{c_d}\,\overline{u} \tag{6.15}$$

and hence the horizontal eddy flux of dry air is small compared to the total horizontal flux.

The problem of estimating $<\overline{S_c}>$ in the presence of horizontal and vertical advection has been addressed recently by Lee (1998), Finnigan (1999), Paw U et al. (2000), Finnigan (1999) and by Massman and Lee (2002). The horizontal flux divergence terms in Equation 6.10 were assumed by Lee (1998) and by Paw U et al. (2000) to be small compared to those in the vertical, but this assumption was shown to be incorrect by Finnigan (1999). He concluded that partial corrections for advection, using the vertical flux divergence terms but neglecting the horizontal terms, were likely to introduce significant error in the estimate of the net exchange between the surface and the atmosphere. Thus both the vertical and horizontal mean flux divergence terms must be considered when calculating the net exchanges of c for air flow over inhomogeneous terrain. Horizontal advection is introduced by inhomogeneity in the flow ($\partial \overline{c_d u}/\partial x \neq 0$) and/or in the source ($\partial \overline{S_c}/\partial x \neq 0 \Rightarrow \partial \overline{\chi}_c/\partial x \neq 0$). Similar considerations apply to vertical advection.

4 Steady, One-dimensional Horizontally Homogeneous Flows

4.1 Fluxes

There is no horizontal advection when the flow is steady and horizontally homogeneous and, because there are no sources of dry air in the control volume, the term $\overline{wc_d}|_h = 0$ in Equation 6.13, i. e. there is no net flux of dry air at height h. This is the key governing constraint used by WPL to develop their theory for correcting eddy covariance measurements for the influence of density fluctuations on trace gas concentration measurements. Under these conditions we can equate the eddy flux density measured at height h to the horizontally averaged source strength, viz.

$$\overline{F}_c = <\overline{c_d}\overline{w'\chi'_c}|_h> = <\overline{S_c}>/A \tag{6.16}$$

where A is the basal area of the control volume. Equation 6.16 shows that the flux density is equal to the product of the mean concentration of dry air and the covariance of vertical velocity and mixing ratio, $\overline{w'\chi'_c}$, measured at height h. Equation 6.16 is identical to that developed by WPL (their Equation 20), except that we have used molar, rather than mass, concentration units to define the mixing ratio. This equation applies to other constituents in the control volume, such as water vapor.

4.2 The vertical velocity of air

Starting from the equation of state $p = cRT$, where p is the total pressure of moist air, R is the ideal gas constant and T is air temperature (°K), we may show that in response to fluctuations in water vapor concentrations, temperature and pressure, fluctuations in the concentration of dry air c'_d are given by

$$c'_d = -c'_v - \bar{c}\left[\frac{T'}{\bar{T}} - \frac{p'}{\bar{p}}\right] \quad (6.17)$$

As discussed above, there is no net flux of dry air through the surfaces of the control volume, and thus in this one-dimensional case

$$\overline{wc_d} = \bar{w}\,\bar{c}_d + \overline{w'c'_d} = 0 \quad (6.18)$$

Combining Equations 6.17 and 6.18, we see that

$$\bar{w} = \frac{1}{\bar{c}_d}\left[\overline{w'c'_v} + \bar{c}\left(\frac{\overline{w'T'}}{\bar{T}} - \frac{\overline{w'p'}}{\bar{p}}\right)\right] \quad (6.19)$$

where we have only retained terms to first order in the fluctuations. This is a much simplified version of that given by Fuehrer and Freihe (2002). The original derivation by WPL did not include the covariance $\overline{w'p'}$, but using a scale analysis, Sun et al. (1995) argued that the $\overline{w'p'}$ term is unimportant relative to the other terms except when heat fluxes are low and wind speeds are high over aerodynamically rough surfaces. At such times the heat flux itself is small and neglect of $\overline{w'p'}$ introduces only small errors in $\bar{w}$. The covariance $\overline{w'p'}$ is expected to be very small compared to the other two terms when there is no asymmetry in the mean static pressure of upward and downward moving eddies (mean pressure is constant and $\partial p/\partial z \simeq 0$ in the surface boundary layer). This contrasts with the asymmetry in density of air where there are net fluxes of sensible and latent heat. A mean vertical velocity of *moist* air arises whenever there are air density fluctuations induced by non-zero fluxes of water vapor or sensible heat. To a high degree of approximation we may thus write

$$\bar{w} = \frac{1}{\bar{c}_d}\left[\overline{w'c'_v} + \bar{c}\frac{\overline{w'T'}}{\bar{T}}\right] \quad (6.20)$$

WPL also derived an expression for $\overline{w}$ (m s^{-1}) in terms of the fluxes of latent heat, λE, and sensible heat, $\overline{H}$. At typical mid-latitude temperatures and pressures

$$\overline{w} = 10^{-6}(0.54\lambda\overline{E} + 2.80\overline{H}) \tag{6.21}$$

where the energy fluxes are in units of W m^{-2}. Under most conditions $\overline{w} < 3$ mm s^{-1}.

Sun et al. (1995) showed that equating $\rho c_p \overline{w'T'}|_h$ with the H at the surface neglects a component of the H associated with the flux of water vapor that occurs when the temperature of the moisture entering the lower surface of the control volume differs from that leaving the upper surface. This term is generally very small and will be ignored here. Similarly, radiative flux divergence between the surface and the measurement height is also neglected in constructing the energy balance.

5 Practical Considerations

5.1 Fluxes in terms of mixing ratios and concentrations

In developing the above equations it has been assumed that concentrations, mixing ratios and velocities can all be measured as required. When concentrations are measured instead of mixing ratios, the flux of constituent c is written as $\overline{F_c} = \overline{w}\,\overline{c}_c + \overline{w'c'_c}$. Combining this with Equation 6.20 for the mean vertical velocity, WPL obtained

$$\overline{F}_c = \overline{c}_d \overline{w'\chi'_c} = \overline{w'c'_c} + \frac{\overline{c}_c}{\overline{c}_d}\left[\overline{w'c'_v} + \overline{c}\frac{\overline{w'T'}}{\overline{T}}\right] \tag{6.22}$$

The two terms on the right correct the eddy flux for the fluctuations in c due to fluctuations in water vapor concentration and temperature when latent heat or sensible fluxes are non-zero. Note that no such corrections are necessary when mean mixing ratios are used to calculate the eddy flux.

Both forms of Equation 6.22 are useful, depending on whether a close-path or open-path analyzer is used to measure the concentrations of the trace constituent and water vapor. We first examine the use of closed-path analyzers to calculate fluxes and then open-path ones.

5.2 Closed-path analyzers

The mixing ratio form of Equation 6.22 is convenient when closed-path gas analyzers are used, thereby eliminating the need to correct for fluxes of water vapor and sensible heat. Thus while the instrument measures concentrations of water vapor, c_v, and CO_2, c_c, the mixing ratio may be calculated, provided temperature and pressure are also measured simultaneously at the sampling frequency used for water vapor and CO_2 (typically 20 Hz). The mixing ratios for water vapor and CO_2 are given by

$$\chi_w = c_v/(c - c_v), \chi_c = c_c/(c - c_v) \qquad (6.23)$$

where $c = p/RT$ is the total molar concentration in the analyzer chamber at any instant. Pressure fluctuations in the air stream are also taken into account through variations in c.

This approach is attractive since there is no need to assume that all temperature fluctuations have been removed from the signal by the time the air travels from the tubing inlet to the analyzer chamber. It is often assumed that perfect temperature equilibrium is achieved at all frequencies contributing to $\overline{F_c}$, allowing the $\overline{w'T'}$ correction term in Equation 6.22 to be set to zero. However, it is unlikely that all the temperature fluctuations will be eliminated at frequencies $\geq 1/(2\pi t_{av})$, where t_{av} is the averaging period (Leuning and Judd 1996). There will then be some unknown residual covariance between w and T, leading to incorrect estimates of the flux. It is thus recommended that the measured trace gas concentrations be converted to mixing ratio in dry air each instant the gas concentration is measured.

Use of Equation 6.23 assumes that the water vapor and CO_2 concentrations are in phase and that they are attenuated by the same amount as the air travels down the tubing. This assumption is needed, irrespective of the way in which concentrations are expressed and the final eddy flux is calculated. The error in χ_c will be small since both fluctuations and absolute values of $c_v \ll c$.

Fluctuations in gas concentrations (and hence mixing ratios) are diminished as air flows through the sampling tubing and gas analyzer (Taylor 1954, Philip 1963) and it is thus necessary to apply corrections to the resultant low-pass filtering (Leuning and Moncrieff 1990, Massman 1991, Lenschow and Raupach 1991, Suyker and Verma 1993, Leuning and Judd 1996). The required corrections can be calculated using theory presented in Leuning and Judd (1996, equations 16-19). Further corrections to loss of covariance resulting from the effects of line averaging and spatial separation between the sonic anemometer and the air inlet to can be calculated using the theory presented by Moore (1986), Leuning and Judd (1996), Massman (2000) and Massman and Lee (2002).

Massman in Chapter 7 states that corrections to the calculated flux due to low-pass filtering need to be applied before the WPL corrections are applied. This is true if concentrations and the rightmost form of Equation 6.22 are used to calculate the flux, but only the corrections for low-pass filtering need be applied when the flux is calculated using mixing ratios relative to dry air.

5.3 Open-path gas analyzers

The problem is different for open path systems because we are unable to calculate the mixing ratio point by point as above. We thus have to apply the WPL corrections involving the fluxes of sensible heat and water vapor.

The terms on the right of Equation 6.22 apply when concentrations are measured in situ using an open-path analyzer. In this case the order in which the fluxes are calculated and the WPL corrections are applied is important. The following steps are recommended

- Calculate the sensible heat flux, $\overline{H}$, according to

$$\overline{H} = \overline{\rho} c_p \overline{w'T'} \tag{6.24}$$

then make corrections for line-averaging along the sonic path length and allow for any separation between the sonic w-axis and the thermometer (e. g., if a separate fine wire or thermocouple is used). Theory presented by Moore (1986), Leuning and Judd (1996) or Massman (2000) may be used to make the required corrections. In Equation 6.24, $\overline{\rho}$ is the mean density of moist air and c_p is the specific heat of air, both in mass units.

Sensible heat fluxes are calculated using Equation 6.24 when temperature fluctuations are measured independently of the vertical wind speed. Most installations use the sonic virtual temperature, defined as $T_s = T(1 + 0.32\chi_v)$ (Kaimal and Gaynor 1991). Thus after Reynolds averaging we have to a close approximation

$$\overline{w'T'} = \overline{w'[T_s/(1 + 0.32\chi_v)]'} \tag{6.25}$$

where the higher order terms in χ_v have been omitted.

- Calculate the flux of water vapor using

$$\overline{E} = (1 + \overline{\chi}_v)[\overline{w'c'_v} + (\overline{c}_v/\overline{T})(\overline{H}/\overline{\rho}c_p)] \tag{6.26}$$

The sensible heat flux has already been corrected for loss of covariance between w and T in step 1, so it is only necessary to correct for loss of covariance between w and c_v. We cannot apply a single correction to both $\overline{w'T'}$ and $\overline{w'c'_v}$ because the geometry will generally differ for the instruments used to measure temperature and water vapor.

- Calculate CO_2 flux. WPL showed that the last two terms on the right of Equation 6.22 may be written in terms of the fluxes of water vapor and sensible heat. Thus for CO_2 we have

$$\overline{F}_c = \overline{w'c'_c} + \overline{c}_c \left[\frac{\overline{E}}{\overline{c}} + \frac{\overline{H}}{\rho c_p \overline{T}} \right] \qquad (6.27)$$

Both sensible heat flux and evaporation have been corrected for loss of covariance in the previous steps, so it is only necessary to correct for loss of covariance between w and c_c due to line averaging and spatial separation of instruments.

Careful experimental design will reduce the magnitude of the latter corrections (Leuning and Moncrieff 1990, Suyker and Verma 1993, Leuning and Judd 1996, Massman and Lee 2002). Instruments should be placed as close together as possible while minimizing flow distortion around the sonic anemometer. Instruments should also be placed as high as possible above the zero-plane displacement height while still remaining within the internal boundary layer of the surface being studied. Loss of covariance can also occur if the averaging period is not sufficiently long to capture the low-frequency contributions to the covariance (Finnigan et al. 2003). These contributions are likely to be site-specific and some analysis will be necessary to determine an adequate averaging period for each experimental site.

5.4 Advection

Most of the above has concentrated on the corrections to the eddy flux of a trace constituent needed to account for density fluctuations induced by the fluxes of water vapor and latent heat. As Equation 6.13 shows, the eddy flux is only one component of four needed to estimate the source term, and $\overline{F}_c = <\overline{c}_c \overline{w'\chi'_c}|_h> = <\overline{S}_c>/A$ only under the restrictive conditions of steady, horizontally homogeneous flows. It is the experience of ourselves, and many other researchers, that the eddy flux provides a poor estimate of the source term when the air flow is stably stratified which often occurs at night. The advection terms in Equation 6.13 then dominate and it is necessary to devise new theoretical and experimental approaches to estimating $<\overline{S}_c>$ under these conditions. This represents a major challenge for the micrometeorological community. A more thorough discussion of advection can be found in Chapter 10.

6 Conclusions

The expressions for mass conservation are relatively simple when concentrations are expressed as molar mixing ratios relative to dry air

(Equation 6.13). This contrasts with the more complex expressions which arise when absolute concentrations are used (e.g. Paw U et al. 2000, Massman and Lee 2002, Fuehrer and Friehe 2002). The mass conservation equation expressed horizontal and vertical advection in terms of mass fluxes of dry air and gradients in mixing ratio, and requires the covariance of vertical wind speed and mixing ratios for the vertical turbulent eddy fluxes. Equation 6.22 shows that the theory developed by WPL for 1-D flows is then still applicable. The right hand side of Equation 6.22 should be used to calculate the vertical eddy flux density when concentrations are measured *in situ*, and the left hand side when a closed-path gas analyzer is employed. In the latter case, measured concentrations should be converted to mixing ratio at the sampling frequency used for eddy covariance. Thus water vapor concentration, temperature and pressure within the gas analysis chamber must be measured simultaneously to calculate the mixing ratio χ_c.

7 References

Baldocchi, D., Finnigan, J. J., Wilson, K., Paw U, K. T., and Falge, E.: 2000, 'On measuring net ecosystem carbon exchange over tall vegetation on complex terrain', *Bound.-Layer Meteorol.* **96**, 257-291.

Baldocchi, D. D., Falge, E., Gu, L. H., Olson, R., Hollinger, D., Running, S., Anthoni, P., Bernhofer, Ch., Davis, K., Evans, R., Fuentes, J., Goldstein, A., Katul, G., Law, B., Lee, X.-L., Malhi, Y., Meyers, T., Munger, W., Oechel, W., Paw U, K. T., Pilegaard, K., Schmid, H. P., Valentini, R., Verma, S., Vesala, T., Wilson, K., and Wofsy, S.: 2001, 'FLUXNET: A new tool to study the temporal and spatial variability of ecosystem-scale carbon dioxide, water vapor and energy flux densities', *Bull. Am. Meteorol. Soc.* **82**, 2415-2434.

Finnigan, J. J.: 2004, 'A re-evaluation of long-term flux measurement techniques. Part 2: Coordinate systems', *Bound.-Layer Meteorol.*, in review.

Finnigan, J. J.: 1999, 'A comment on the paper by Lee (1988): "On micrometeorological observations of surface-air exchange over tall vegetation"', *Agric. For. Meteorol.* **97**, 55-64.

Finnigan, J. J., Clements, R., Malhi, Y., Leuning, R., and Cleugh, H. A.: 2003, 'A re-evaluation of long-term flux measurement techniques. Part 1: Averaging and coordinate rotation', *Bound.-Layer Meteorol.* **107**, 1-48.

Fuerher, P. L. and Friehe, C. A.: 2002, 'Flux corrections revisited', *Bound.-Layer Meteorol.* **102**, 415-457.

Kaimal, J. C. and Gaynor, J. E.: 1991, 'Another look at sonic thermometry', *Bound.-Layer Meteorol.* **56**, 401-410.

Kramm, G., Dlugi, R., and Lenschow, D. H.: 1995, 'A re-evaluation of the Webb correction using density-weighted averages', *J. Hydrol.* **166**, 283-292.

Lee, X.: 1998, 'On micrometeorological observations of surface-air exchanges over tall vegetation', *Agric. For. Meteorol.* **91**, 39-49.

Lenschow, D. H. and Raupach, M. R.: 1991, 'The attenuation of fluctuations in scalar concentrations through sampling tubes', *J. Geophys. Res.* **96**, 5259-5268.

Leuning, R. and Judd, M. J.: 1996, 'The relative merits of open- and closed-path analyzers for measurements of eddy fluxes', *Global Change Biology* **2**, 241-253.

Leuning, R. and Moncrieff, J.: 1990, 'Eddy-covariance CO_2 flux measurements using open-path and closed-path CO_2 analyzers - Corrections for analyzer water vapor sensitivity and damping of fluctuations in air sampling tubes', *Bound.-Layer Meteorol.* **53**, 63-76.

Massman, W. J.: 1991, 'The attenuation of concentration fluctuations in turbulent-flow through a tube', *J. Geophys. Res.* **96**, 5269-5273.

Massman, W. J.: 2000, 'A simple method for estimating frequency response corrections for eddy covariance systems', *Agric. For. Meteorol.* **104**, 185-198.

Massman, W. J. and Lee, X.: 2002, 'Eddy covariance flux corrections and uncertainties in long-term studies of carbon and energy exchanges', *Agric. For. Meteorol.* **113**, 121-144.

McMillen, R. T.: 1988, 'An eddy correlation technique with extended applicability to non-simple terrain', *Bound.-Layer Meteorol.* **43**, 231-245.

Moore, C. J.: 1986, 'Frequency response corrections for eddy correlation systems', *Bound.-Layer Meteorol.* **37**, 17-35.

Paw U, K. T., Baldocchi, D. D., Meyers, T. P., and Wilson, K. B.: 2000, 'Correction of eddy-covariance measurements incorporating both advective effects and density fluxes', *Bound.-Layer Meteorol.* **97**, 487-511.

Philip, J. R.: 1963, 'The damping of a fluctuating concentration by continuous sampling through a tube', *Australian J. Physics* **16**, 454-463.

Sun, J. Esbensen, S. K. and Mahrt, L.: 1995, 'Estimation of surface heat flux', *J. Atmos. Sci.* **52**, 3162-3171.

Suyker, A. E. and Verma, S. B.: 1993, 'Eddy correlation measurement of CO_2 flux using a closed-path sensor: Theory and field tests against open-path sensor', *Bound.-Layer Meteorol.* **64**, 391-407.

Taylor, G. I.: 1954, 'The dispersion of matter in turbulent flow through a pipe', *Proc. R. Soc. London* **A223**, 446-468.

Webb, E. K., Pearman, G. I., and Leuning, R.: 1980, 'Correction of flux measurements for density effects due to heat and water vapor transfer', *Quart. J. R. Meteorol. Soc.* **106**, 85-100.

Wilczak, J. M., Oncley, S. P., and Stage, S. A.: 2001, 'Sonic anemometer tilt correction algorithms', *Bound.-Layer Meteorol.* **99**, 127-150.

Chapter 7

CONCERNING THE MEASUREMENT OF ATMOSPHERIC TRACE GAS FLUXES WITH OPEN- AND CLOSED-PATH EDDY COVARIANCE SYSTEM: THE WPL TERMS AND SPECTRAL ATTENUATION

William Massman
wmassman@fs.fed.us

Abstract

Atmospheric trace gas fluxes measured with an eddy covariance sensor that detects a constituent's density fluctuations within the *in situ* air need to include terms resulting from concurrent heat and moisture fluxes, the so called 'density' or 'WPL corrections' (Webb et al. 1980). The theory behind these additional terms is well established. But, virtually no studies to date have examined the constraints imposed on the theory by different instrumentation technologies and by limitations inherent to eddy covariance systems. This study extends the original WPL theory by examining how eddy covariance instrumentation, particularly spectral attenuation and an instrument's basic technology, influences the application of this theory to flux measurement. Specific issues discussed here include the importance of static pressure fluctuations to the WPL theory, the possible systematic overestimation of the WPL vapor term, and the transfer functions associated with signal processing and volume averaging effects of a fast-response closed-path CO_2/H_2O sensor. This different perspective on the WPL theory suggests that current methods of applying the WPL theory, particularly with closed-path systems, can yield significant biases in the annual carbon balance derived from eddy covariance technology and can cause the surface energy imbalance to increase with increasing wind speed. Furthermore, it is suggested that spectral corrections should be made before applying the WPL theory to estimate fluxes and that high frequency point-by-point conversions from mass density to mixing ratio is not the preferred method for estimating fluxes by eddy covariance.

1 Introduction

Webb et al. (1980), henceforth WPL80, showed that eddy covariance trace gas fluxes measured with a sensor that detects a constituent's density fluctuations within the *in situ* air need to include terms resulting from concurrent heat and moisture fluxes. These additional terms arise as a consequence of the density fluctuations of the ambient air sampled by an instrument that measures trace gas density rather than the constituent's molar mixing ratio (WPL80; Paw U et al. 2000, Fuehrer and Friehe 2002, Massman and Lee 2002). Unfortunately, so far no technology has been developed that allows a single instrument to directly sense a constituent's mixing ratio. So measured mass fluxes will continue to require additional instrumentation for heat and moisture fluxes.

Since WPL80 this theory has been validated for an open-path eddy covariance system (e. g., Leuning et al. 1982), developed and compared for open- and closed-path systems (Leuning and Moncrieff 1990, Leuning and King 1992, Suyker and Verma 1993, Lee et al. 1994, Leuning and Judd 1996), extended to include other terms, most notably the fluctuating pressure term, (e. g., Fuehrer and Friehe 2002, Massman and Lee 2002), and redeveloped in three dimensions (Paw U et al. 2000, Massman and Lee 2002), and further refined in Chapter 6.

In general there is little doubt about the validity or appropriateness of this theory. However, much of the discussion and development of this theory to date has centered on applying it to different types of eddy covariance instruments, i. e., to open- and closed-path systems. This study takes a different approach by examining how the instrumentation, particularly spectral attenuation and an instrument's basic technology, influences the application of the WPL80 theory to the measurement of eddy covariance fluxes. Central to this issue are the questions of whether spectral corrections should be made before or after applying the WPL80 theory to estimate fluxes and whether making high frequency point-by-point conversions from mass density to mixing ratio is useful for estimating fluxes. Some of these issues have been (at least partially) addressed in previous work and some have not.

To accomplish this goal three fundamentals need to be presented. First, in this study the terms flux and covariance are not used synonymously. Here flux refers to mass transfer rates in the atmosphere. Covariance, on the other hand, refers to the covariance between signals, or truncated data streams, obtained by two different instruments. Thus covariances are associated with instruments. Furthermore, it is assumed here that no eddy covariance instrument is necessarily free of high frequency attenuation and that the amount of attenuation can be unique to any given instrument or eddy covariance system. In most cases it is generally assumed that correcting the covariances for spectral attenuation yields an estimate of the flux. However, as discussed later, this is not necessarily the case for a closed-path system. Therefore, a distinction is also made between corrected and uncorrected covariances.

Second, the WPL80 terms are not a consequence of inadequate sensor performance, and in that sense they are not instrument related corrections. Any properly functioning CO_2 instrument that employs infrared gas analysis technology detects the number of absorbing CO_2 molecules within the path of its infrared light beam. Assuming that an instrument detection volume is constant, then a CO_2 instrument indirectly measures the density (or number density) of the CO_2 molecules in a sample. Consequently, the WPL80 temperature, pressure, and vapor terms are not required to 'correct' the measured trace gas density—Fuehrer and Friehe (2002) make the same point. Rather they are required to compensate for the concurrent density fluctuations associated with fluctuations in temperature, water vapor, and pressure in the air sampled with this type of instrument. In essence the WPL80 terms are required to distinguish between the true surface exchange (or biologically relevant) flux and the atmospheric flux measured with a sensor that detects mass density rather than mixing ratio. As a result this study will not refer to the WPL80 terms as corrections.

Third, in principle the WPL80 terms apply to (or characterize) the ambient environment in which the trace gas density is measured. For example, the ambient environment at the place of measurement in a closed-path system is not characterized by high frequency temperature fluctuations because the intake tube attenuates the temperature fluctuations so strongly that they can be ignored (Frost 1981, Leuning and Moncrieff 1992, Rannik et al. 1997). [As discussed later, this assumption, although valid at high frequencies, may not be valid at low frequencies (Leuning 2003, personal communication).] In effect, therefore, the intake tube alters the sample used to measure the atmospheric trace gas density. This ability to alter the measurement sample is a crucial difference between open- and closed-path sensors. Both open- and closed-path sensors are similar in that they include an infrared gas analyzer that responds to the attenuation (by absorption) of an infrared light beam. However, they are fundamentally different in their sampling strategy because the open-path system is a passive system (i. e., it does not fundamentally alter the measurement sample), whereas the closed-path system is an active system because it does alter the sample. When estimating fluxes this distinction is critical to the application of spectral corrections to the covariances and the WPL80 terms.

The intent of this study is to systematically examine open- and closed-path systems by applying the above three fundamentals to each in turn. Consequently, this study also examines the transfer functions appropriate to the signal processing software and volume averaging effects of a closed-path instrument, as well as, possible influences that the pressure fluctuations can have on fluxes measured with a closed-path system. The next section formulates the relationship between the flux, the WPL80 (temperature, pressure, and vapor) terms, and spectral attenuation. After that sections 3 and 4 discuss open- and closed-path systems, with section 4 presenting some new aspects of closed-path systems. The final section of this study summarizes the conclusions.

2   The WPL80 Terms and Spectral Attenuation

The turbulent atmospheric mass flux of a trace gas measured with an instrument that measures the mass mixing ratio of the gas (ω_g) rather than its density (ρ_g) is expressed as $\overline{\rho}_d \overline{w'\omega'_g}$; where the overbar is the time averaging or covariance operator, $\overline{\rho}_d$ is the time-averaged (mean) dry air density, w' is the fluctuating vertical velocity, and ω'_g is the fluctuation of the trace gas mass mixing ratio (ω_g), where ω_g is defined as the ratio of the trace gas density to the dry air density: $\omega_g = \rho_g/\rho_d$. WPL80 developed the following relationship between $\overline{\rho}_d \overline{w'\omega'_g}$ and the heat, pressure, and mass fluxes measured with instruments that detect changes in density rather than mixing ratio:

$$\overline{\rho}_d \overline{w'\omega'_g} = \overline{w'\rho'_g} + \overline{\rho}_g(1+\overline{\chi}_v)\left[\frac{\overline{w'T'_a}}{\overline{T}_a} - \frac{\overline{w'p'_a}}{\overline{p}_a}\right] + \mu_v \overline{\omega}_g \overline{w'\rho'_v} \qquad (7.1)$$

where ρ'_g is the trace gas density fluctuation, $\overline{\chi}_v$ is the mean volume mixing ratio for water vapor (which is the ratio of mean vapor pressure, $\overline{p}_v$, to the mean partial pressure of the dry air, $\overline{p}_d$; $\overline{\chi}_v = \overline{p}_v/\overline{p}_d$), $\overline{T}_a$ is the mean ambient temperature, T'_a is the fluctuation in ambient temperature, $\overline{p}_a$ is the mean ambient pressure, p'_a is the fluctuation in ambient pressure, μ_v is the ratio of the molecular mass of dry air, m_d, to the molecular mass of water vapor, m_v, (i.e., $\mu_v = m_d/m_v$), and ρ'_v is the fluctuation in the ambient water vapor density. The first term on the right hand side (RHS) of Equation 7.1, $\overline{w'\rho'_g}$, is the density covariance. The second term includes the temperature covariance, $\overline{\rho}_g(1+\overline{\chi}_v)[\overline{w'T'_a}/\overline{T}_a]$, and the pressure covariance, $\overline{\rho}_g(1+\overline{\chi}_v)[-\overline{w'p'_a}/\overline{p}_a]$. The $\mu_v \overline{\omega}_g \overline{w'\rho'_v}$ term is the water vapor covariance.

Although Equation 7.1 is fairly standard there are several associated issues that should be mentioned. First, of the four covariances comprising the RHS of Equation 7.1 only the last three are WPL80 terms and only they are associated with fluctuations in the ambient environment at the point of the measurement of the trace gas density, ρ_g. Second, strictly speaking, WPL80 did not include the pressure flux term in their development although they were aware of it. This term is included here because it has been shown to be important for open-path systems for some atmospheric conditions (Massman and Lee 2002) and because it is needed in order to assess its importance to closed-path systems. Third, the subscript 'g' is used in Equation 7.1 and throughout this study to denote any general trace gas. Carbon dioxide is specified with a 'c' subscript and water vapor is specified by a 'v' subscript.

Equation 7.1 basically assumes that all instruments make perfect measurements (no high frequency attenuation, immediate response, high signal to noise ratios), that such instruments are co-located at a point in space and make simultaneous measurements (no spatial separation or time lag effects), and that the data archiving system is perfect (no digitization noise, no external electronic contamination of the signal, perfect signal processing). In this case the three WPL80 covariance terms are true atmospheric fluxes, the density covariance term, $\overline{w'\rho'_g}$, is the true atmospheric mass flux measured with an instrument that detects fluctuations in mass density rather than mixing ratio, and $\overline{\rho}_d \overline{w'\omega'_g}$ is the true surface exchange flux. Of course no such system exists and all quantities and covariances measured in Equation 7.1 are compromised somewhat. Thus the measured (or more properly the uncorrected) surface flux, $(\overline{\rho}_d \overline{w'\omega'_g})_m$ is better represented by

$$(\overline{\rho}_d \overline{w'\omega'_g})_m = (A^d_{wg} \overline{w'\rho'_g}) + \overline{\rho}_g (1 + \overline{\chi}_v) \left[\frac{(A_{wT} \overline{w'T'_a})}{\overline{T}_a} - \frac{(A_{wp} \overline{w'p'_a})}{\overline{p}_a} \right] + \mu_v \overline{\omega}_g (A_{wv} \overline{w'\rho'_v}) \quad (7.2)$$

where the subscripted A is an attenuation factor that represents the aggregated instrument and system related effects that tend to reduce the true covariance (i. e., $0 \leq A \leq 1$). For this study each of the four terms on the RHS of Equation 7.2 represents an uncorrected covariance between the vertical velocity and another instrument and the subscript attached to each attenuation factor identifies a particular covariance. The 'd' superscript on the first term is to distinguish between the attenuation factor for the density term and those associated with the WPL80 terms. This last distinction is important for closed-path systems.

Correcting these covariances for spectral attenuation has been the subject of many recent studies (see the following papers and their references for a summary: Massman 2000 and 2001, Rannik 2001, Chapter 4). This study is similar to these previous studies in that it also develops some new transfer functions appropriate to (at least some) closed-path eddy covariances systems. These transfer functions are based on filters implemented as part of the signal processing software and the volume averaging effects of the sampling chamber. But this study also extends the previous studies of spectral corrections by placing them in the context of the WPL80 terms as they relate to open- and closed-path systems. In essence the next two sections address the steps required to derive a corrected flux estimate, $\overline{\rho_d w' \omega_g'}$, from an uncorrected flux estimate, $(\overline{\rho_d w' \omega_g'})_m$ for open- and closed-path systems.

3 Open-path Systems

Both open- and closed-path systems produce attenuated signals. However, attenuation of CO_2 or H_2O density fluctuations in an open-path sensor results from the sensor's inability to resolve data on scales smaller than the detection volume. This is an instrument design issue and is not related to actively altering the sample's temperature, pressure, water vapor, or CO_2 content. In the case of a closed-path sensor the intake tube physically attenuates the temperature fluctuations and the CO_2 and H_2O density fluctuations by mechanical mixing, molecular diffusion, and interaction with the tube walls. It can also both enhance and attenuate the pressure fluctuations (Iberall 1950, Holman 2001). Beyond these tube effects the instrument itself (e. g., a Licor 6262 or other closed-path instrument) also attenuates the signal. Some of this attenuation is flow-related and is similar to the tube effects. Some of it is related to volume averaging and signal processing, which like the open-path sensor do not physically alter the sample. Only the flow path actively (although possibly inadvertently) acts to alter the sample by changing its temperature, damping its moisture and CO_2 variations, and altering its pressure fluctuations.

Strictly speaking there are several (albeit relatively minor) reasons why an open-path sensor is not a truly passive sensor. For example, the energy of the infrared signal absorbed by the CO_2 molecules increases their vibrational and rotational energy (a quantum physical effect). In addition, the sensor can actually remove mass from the sample when condensation occurs on the lenses, which generally causes an easily diagnosed problem by rendering the data useless. There are also the possibilities that the sensor may distort the flow and that there are boundary-layer effects associated with flow near the flat surfaces that enclose the optical path. Further, open-path sensors are a heat source to the atmosphere because of their infrared signal generator and because (and maybe more importantly) the sensor body radiates absorbed solar radiation as heat. Conceivably, these last two effects could alter the temperature of the sample before or during its passage through the instrument's optical path. However, these issues and all the previous effects can be ignored for the present discussion.

An open-path sensor is intended to be used in the open atmosphere. It is in that sense an *in situ* sensor. Therefore, it samples the ambient environmental conditions and all the WPL80 terms are associated with the ambient environment. Consequently all the covariances in Equation 7.2 are related to atmospheric fluxes. Furthermore, all uncorrected covariances measured with an open-path system must be spectrally corrected before summing them to produce an estimate of $\overline{\rho}_d \overline{w'\omega'_g}$. A simple thought experiment should help to clarify this issue. Consider two cases for measuring the surface CO_2 flux. The first case is for the perfect instrument or system, for which no spectral corrections apply; i. e., all instruments are co-located and perfectly measure data at a point so that $A^d_{wc} = A_{wT} = A_{wp} = A_{wv} = 1$ and $(\overline{\rho}_d \overline{w'\omega'_c})_m = \overline{\rho}_d \overline{w'\omega'_c}$. In this case the WPL80 terms are simply added to the density covariance term, $\overline{w'\rho'_c}$, to yield the true CO_2 surface mass flux, $\overline{\rho}_d \overline{w'\omega'_c}$.

The second case differs from the first only in that the CO_2 measurement is attenuated by 25% (i. e., $A^d_{wc} = 0.75$ and $A_{wT} = A_{wp} = A_{wv} = 1$). For this example the only way to recover the true surface flux, $\overline{\rho}_d \overline{w'\omega'_c}$, from the uncorrected surface flux, $(\overline{\rho}_d \overline{w'\omega'_c})_m$ is to correct the attenuated density covariance, $A^d_{wc} \overline{w'\rho'_c}$, then add all the WPL80 terms to it. Applying the spectral corrections after including the WPL80 terms would be equivalent to correcting $(\overline{\rho}_d \overline{w'\omega'_c})_m$ directly, which in turn would also multiply (or over-correct) the three WPL80 terms by a correction factor that applies only to the CO_2 instrument. This could yield a significantly biased estimate of the true surface flux because for most environments the WPL80 temperature covariance term is often the dominant term. This example can be extended to include any combination of imperfect (spectrally attenuated) covariance measurements and in general one must conclude that for open-path sensors spectral corrections must be applied to the uncorrected covariances before including the WPL80 terms in the final estimate of the trace gas surface flux. The only exceptions to this are the very unlikely situations where either all covariance attenuation factors are identical or all WPL80 terms are negligibly small compared with the density covariance term. In general it must be assumed that spectral (or cospectral) corrections are specific to the instruments involved and that they are not necessarily transferable from one covariance system to another. In other words, individual instruments are often based on fundamentally different technologies, which can impose different physical designs and separation distances, different time constants, and different noise reducing filters. All of these define instrument specific response functions.

This basic principle of instrumentation and the other fundamentals, discussed previously, are also relevant to closed-path systems and to the estimation of surface fluxes by converting high frequency CO_2 mass density measurements (ρ'_c) to high frequency CO_2 dry-air mass mixing ratio (ω'_c).

4 Closed-path Systems

4.1 General considerations

A closed-path system is a combination of both active and passive sampling. Attenuation of the temperature fluctuations in a closed-path system qualifies as active because it results from a combination of molecular and turbulent diffusion within the intake tube and the associated heat exchange with the tube walls. In essence the tube acts as a heat exchanger and alters the sample temperature before it is drawn into the detection chamber of the infrared gas analyzer.

Attenuation of fluctuations in trace gas mass density result from a combination of diffusional smoothing of density variations inside the flow path (defined by the tube and the detection chamber), possible absorption/adsorption/desorption of the trace gas to or from the walls of the flow path, design (line or volume averaging) aspects of the infrared gas analyzer's detection chamber, and any signal processing or electronic filtering inherent to the instrument's electronic circuitry. Of these only the tube and chamber flow effects qualify as active, all others are passive.

Usually, however, these active and passive effects are lumped together into a single time constant, which is then used to describe the closed-path system. But, including the WPL80 terms in a manner appropriate to a closed-path system requires careful consideration of the nature of the sampling and its associated spectral correction. In general the spectral corrections made to the WPL80 covariance terms should *not* include any active (or flow path) attenuation effects. Rather, they should include only passive attenuation effects associated with the other parts of the system. This may seem surprising at first, but it follows directly from the fact that the WPL80 covariance terms refer to the environment in which the trace gas density is measured. Therefore, the appropriate measure of p'_a in the WPL80 pressure covariance term, $\overline{\rho}_g(1+\overline{\chi}_v)[-\overline{w'p'_a}/\overline{p}_a]$, and of ρ'_v in the WPL80 vapor covariance term, $\mu_v\overline{\omega}_g\overline{w'\rho'_v}$, are those occuring within the detection chamber of the closed-path system. This result follows from the same logic (or physical manipulation of the sample) that eliminates the high frequency T'_a (and the associated WPL80 temperature covariance term) from the environment of the detection chamber. It's just that in the case of T'_a the attenuation is usually considered 100% effective, but for p'_a and ρ'_v the physical attenuation of the signals is not as complete. (Note that the attenuation of p'_a by the flow path is made more precise later in this study.)

This result, which applies to both CO_2 and water vapor because they are often measured with the same closed-path system, has some surprising implications for physical interpretation of the WPL80 terms and for estimating and correcting the attenuation factors included in Equation 7.2. For CO_2 the spectral correction factor $(1/A^d_{wc})$ for the density covariance, $A^d_{wc}\overline{w'\rho'_c}$, must exceed the correction factor $(1/A_{wv})$ for the vapor covariance, $A_{wv}\overline{w'\rho'_v}$, because both CO_2 and water vapor covariances share exactly the same set of passive attenuation factors, but only the CO_2 density term includes the active (flow path) attenuation effects as well (i.e., $A^d_{wc} < A_{wv}$). If the active portion of all the attenuation factors is included when spectrally correcting the WPL80 vapor covariance, then the surface CO_2 flux, $\overline{\rho}_d \overline{w'\omega'_c}$, will be overestimated as a result. The same applies for water vapor as well, therefore (and even more surprisingly) $A^d_{wv} < A_{wv}$. In other words, when measuring the water vapor covariance $\overline{w'\rho'_v}$ with a closed-path system the attenuation (or correction) factor that applies to the density covariance is different than the one that applies to the WPL80 term even though they are the same measured quantity.

The reason for this surprising difference is that the density covariance should be interpreted in terms of an atmospheric- or surface-related flux (which is external to the environment in which the measurements are made in a closed-path sensor), whereas the WPL80 terms refer to conditions internal to the instrument. Thus the WPL80 terms lose their interpretations as surface exchange fluxes. Rather they are simply covariances between the sonic anemometer and measurements made inside the chamber of a closed-path system. This is very different from the open-path case for which the WPL80 terms retain their interpretation as surface-related fluxes. But because the closed-path system actively alters the sample the WPL80 terms lose their immediate association with surface fluxes.

Given this distinction between the density covariance and the WPL80 covariance terms and its importance to spectral corrections and the estimation of surface fluxes, the next section develops the transfer functions for the detection chamber of a closed-path system.

4.2 Detection chamber transfer functions

There are two aspects of the closed-path detection chamber that compromise its ability to produce a precise estimate of ρ'_c and ρ'_v within the chamber: the signal processing software and the volume averaging effects of the chamber. Both of these issues are appropriate to open-path instruments as well. But, for open-path instruments the signal processing software effects are usually minimal and can typically be ignored and the volume effects (and related line averaging) have been discussed (Gurvich 1962, Silverman 1968, Andreas 1981, Moore 1986, Massman 2000).

For this study the Licor 6262 is used as an example of how to address these concerns. However, the general approach for developing these transfer functions (if not the specific transfer functions themselves) applies to any closed-path sensor. The next three subsections provide a detailed discussion.

4.2.1 The signal processing algorithm

To provide a good signal to noise ratio the Licor 6262 uses a third-order Bessel filter as an antialiasing filter. Its associated (complex) transfer function, $h_{3B}(\omega)$, is given as follows:

$$h_{3B}(\omega) = \frac{15}{(15 - 6\Omega^2) - j(15\Omega - \Omega^3)} \qquad (7.3)$$

where $j = \sqrt{-1}$, $\Omega = \omega \tau_{3B}\sqrt{3.0824}/(2\pi)$, τ_{3B} [s] is the time constant of the third order Bessel filter, $\omega = 2\pi f$ [radians s^{-1}] and f [Hz] is frequency. Because this filter is complex there is both a real part, the gain function or $H_{3B}(\omega)$, and an imaginary part or phase function, $H^{\phi}_{3B}(\omega)$. The gain function is expressed as

$$H_{3B}(\omega) = \frac{15}{\sqrt{\Omega^6 + 6\Omega^4 + 45\Omega^2 + 225}} \qquad (7.4)$$

The importance of the phase function (to first order instruments) was pointed out by Hicks (1972) and Horst (1997) and further developed to include the effects of any longitudinal displacement between the sonic and the mouth of the intake tube and any possible (unresolved) tube lag times by Massman (2000). Following Massman (2000), the phase function for the third-order Bessel filter appropriate to the present example is

$$H^{\phi}_{3B}(\omega) = \frac{(15 - 6\Omega^2)\cos[\phi(\omega)] - (15\Omega - \Omega^3)\sin[\phi(\omega)]}{\sqrt{\Omega^6 + 6\Omega^4 + 45\Omega^2 + 225}} \qquad (7.5)$$

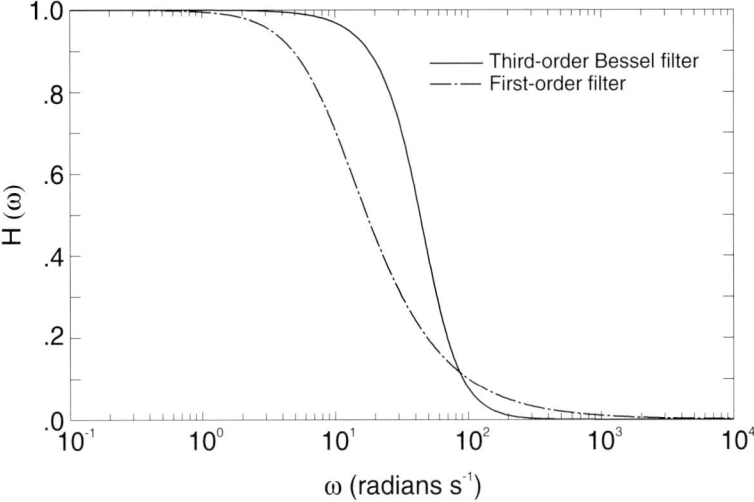

Figure 7.1. Gain functions, $H(\omega)$, for a first order filter with a time constant $\tau_1 = 0.1$ s and third-order Bessel filter with a time constant $\tau_{3B} = 0.2$ s. Equation 7.6 is the first order filter's gain function and Equation 7.4 is for the third-order Bessel filter.

where $\phi(\omega) = \omega(l_{lon}/u + L_t/U_t)$ and l_{lon} is the longitudinal displacement, u is the mean horizontal atmospheric wind speed, L_t is the tube length, and U_t is the tube flow velocity. In most applications the phase effects associated with the tube lag time, L_t/U_t, are eliminated from Equation 7.5 by digitally shifting sonic time series so that it will be synchronized with the closed-path sensor. However, depending on the sampling frequency and the exact value of the lag time, some unresolved lag time may still remain as part of the phase. Here L_t/U_t is included for completeness and will be understood as any possible unresolved tube lag time.

It is possible to compare each of these last two transfer functions with their first order counterparts, $H_1(\omega)$ and $H_1^\phi(\omega)$, which is done by the next two equations and Figures 7.1 and 7.2.

$$H_1(\omega) = \frac{1}{\sqrt{1 + \omega^2 \tau_1^2}} \qquad (7.6)$$

$$H_1^\phi(\omega) = \frac{\cos[\phi(\omega)] - \omega \tau_1 \sin[\phi(\omega)]}{\sqrt{1 + \omega^2 \tau_1^2}} \qquad (7.7)$$

where τ_1 is the time constant of the first order instrument. [Note that these last two equations are expressed differently by Massman (2000),

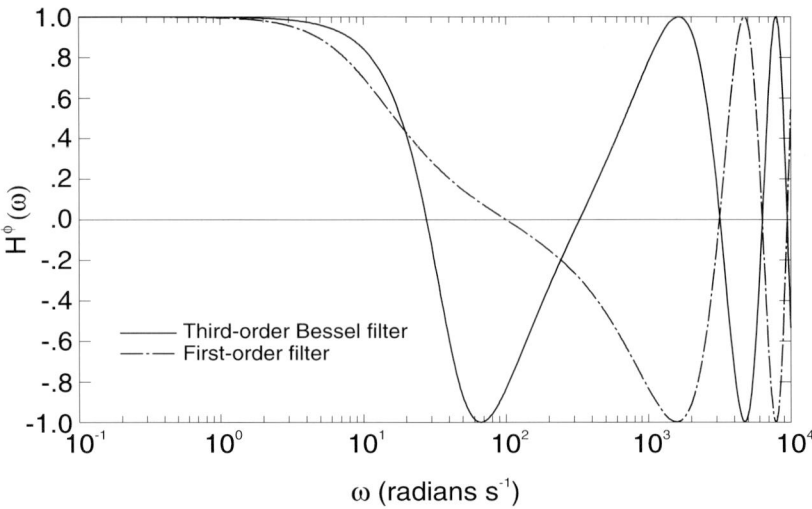

Figure 7.2. Phase functions, $H^\phi(\omega)$, for a first order filter with a time constant $\tau_1 = 0.1$ s and third-order Bessel filter with a time constant $\tau_{3B} = 0.2$ s. Equation 7.7 is the phase function for first order filter and Equation 7.5 is for the third-order Bessel filter. The phase $\phi(\omega) = 0.001\omega$ for both filters.

but that their multiplicative effect for spectral correction factors is the same regardless.]

For the purposes of comparisons only, Figures 7.1 and 7.2 assume that $\tau_1 = 0.1$ s and that $l_{lon}/u + L_t/U_t = 0.001$ s. For the Licor 6262 $\tau_{3B} = 0.2$ s with 0.1 s being its recommended nominal first order equivalent time constant. As indicated in both Figures 7.1 and 7.2 the third-order Bessel filter (with $\tau_{3B} = 0.2$ s) produces less filtering than (or out performs) the first order filter (with $\tau_1 = 0.1$ s). In the case of the phase functions Figure 7.2 indicates that each filter has a different effect on the phase at high frequencies ($\omega \geq 20$ radians s^{-1} or $f \geq 4$ Hz). However, the phase-shifting portions of these filters occur in the cospectral region with very little power so that this behavior is not particularly significant to spectral correction factors or observed cospectra.

4.2.2 Spatial averaging of the detection chamber

The Licor 6262 detection chamber is approximately 0.15 m long, 0.0063 m high, and 0.0126 m wide. The volume flow through the detection chamber and the infrared signal path are parallel and down the length of chamber. The light beam tapers somewhat between one end of the sample chamber and the other, but this will be neglected for the present discussion. Also neglected here is any flow path (active) attenuation of mass fluctuations associated with the detection chamber itself. This is justifiable because the tube length is usually much greater than the length of the detection chamber.

The rectangular geometry of the detection chamber suggests the use of Cartesian coordinates for modeling the volume averaging effects of the sample chamber. It is possible to show, but will not be done here, that this approach is formally or mathematically the same as those used to express the effects of line averaging by open-path sensors on the measured spectra (Gurvich 1962, Silverman 1968). However, there is one important difference. The flow velocity within the detection chamber can be very different than the wind speed of the ambient atmosphere near the tube mouth, so that the transfer functions need to be expressed in terms of the volume flushing time constant of the detection chamber, τ_{vol}, rather than averaging lengths. Therefore, for an instrument with a flow path that is parallel to the infrared light beam the spectral transfer function associated with volume averaging, $H_{vol}(\omega)$, is

$$H_{vol}(\omega) = \frac{\sin^2(\omega \tau_{vol}/2)}{(\omega \tau_{vol}/2)^2} \qquad (7.8)$$

Given the maximum flow rate of the Licor 6262 is 10 L min^{-1} and that the volume of the detection chamber is 0.0119 L, then the minimum value that τ_{vol} that can be expected is about 0.07 s (i.e., $\tau_{vol} \geq 0.07$ s).

Although it is reasonable to assume that the infrared light beam is parallel to the flow path, it is possible that they could deviate slightly from one another. But, it is also possible to account for these deviations. For example, Gurvich (1962) developed the appropriate transfer function for the perpendicular case and Silverman (1968) generalized the Gurvich function to any angle less than 90 degrees. However, these deviations are expected to be small for the 6262 and they will not be investigated here.

4.2.3 Is a closed-path sensor a first order instrument?

The nominal (first order) time constant for the Licor 6262 sensor is often taken to be 0.1 s. This presumption is now tested with a simple example by calculating the spectral correction factors for a first order sensor and a sensor that combines the effects of the third-order Bessel filter with volume averaging. These calculations are performed using the integration approach summarized by Equation 3 of Massman (2000) or Equation 1 of Chapter 4. Here the focus is on the correction factor rather than the transfer functions because, first, the results and conclusions are the same regardless and, second, a practical example using

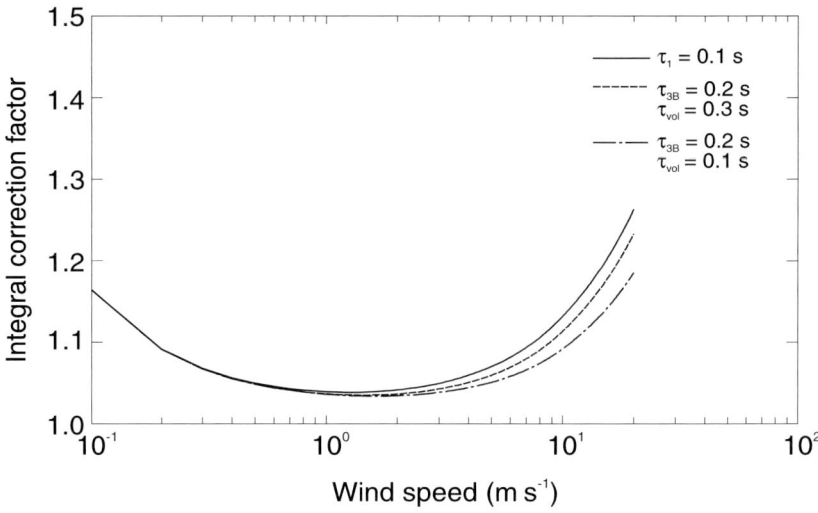

Figure 7.3. Comparison of integral correction factors for three different Licor 6262 scenarios. The first two scenarios combine the effects of the third-order Bessel filter, which is part of the instrument's signal processing software, with the volume averaging effects of the detection chamber. Two different values for the volume flushing time constant, τ_{vol}, are shown. The third scenario assumes that the Licor 6262 is a first order instrument with a response time, τ_1, of 0.1 s. Neutral atmospheric stability is assumed.

correction factors is more insightful for this comparison. All further closed-path scenarios assume the following: (*i*) the height of the covariance measurement is 5 m above the zero plane displacement, (*ii*) the sampling rate is 10 Hz and the sampling period is 30 minutes, (*iii*) the atmosphere is neutrally stable, (*iv*) the sonic path length is 0.15 m, (*v*) the mouth of the intake tube is displaced both laterally and longitudinally by 0.15 m from the center of the sonic, (*vi*) flow in the tube is turbulent and the corresponding Reynolds number is about 20,000, (*vii*) the ratio [including Massman's (1991) tube attenuation factor, Λ] of the tube radius to its length is 0.03 and the total tube lag time (not the unresolved portion) is 2.0 s, and (*viii*) the maximum nondimensional frequency of the frequency-weighted cospectrum, η_x (Massman 2000), is 0.085 after the flat terrain cospectrum of Kaimal et al. (1972), (*ix*) the high pass block averaging filter is included in the calculations of the spectral correction factor, and (*x*) the shape of the cospectrum is taken from Chapter 4.

The results, shown in Figure 7.3, indicate (a) that describing the Licor 6262 as a first order instrument with a time constant of 0.1 s overpredicts the true attenuation somewhat at wind speeds greater than about 3 m s^{-1} and therefore, overpredicts the spectral correction factor for these wind speeds and (b) that the volume averaging effects of the Licor 6262 detection chamber, although relatively small, can contribute to spectral attenuation. Regarding (a), some trial and error comparisons suggested that the Licor 6262 was better described as a first order instrument with a time constant of 0.06 to 0.08 s, depending on τ_{vol}. Result (b) is, of course, somewhat dependent upon the exact values of τ_{vol} and η_x. Larger values for either of these parameters will increase the spectral correction factor.

Part of the reason for (b) is that the appropriate transfer function is actually $\sqrt{H_{vol}(\omega)}$ (e. g., Moore 1986), rather that $H_{vol}(\omega)$ itself, which applies to spectra rather than ρ'_c or ρ'_v. This will tend to reduce the attenuation that would have otherwise have been predicted by $H_{vol}(\omega)$. But, this also highlights an important aspect of making spectral corrections, which is that the assumptions made when deriving a transfer function also determine how it is applied. For example, if the transfer function is developed on the basis of spectra, then taking the square root is appropriate to describe the attenuation of fluctuations. This is usually the case for line averaging or volume averaging effects (e. g., Andreas 1981). However, if the transfer function is derived directly on the basis of mass density fluctuations then taking the square root is not appropriate. A good example of this last case is the transfer function describing tube attenuation effects (e. g., Massman 1991).

4.3 Pressure fluctuations within the detection chamber

For most atmospheric conditions the variations in ambient density due to the pressure covariance term, $\bar{\rho}_g(1 + \bar{\chi}_v)[-\overline{w'p'_a}/\bar{p}_a]$, can be ignored. However, for windy, turbulent conditions and open-path sensors this may not be true (Massman and Lee 2002). It is, therefore, worthwhile to explore the possible nature of the pressure fluctuations inside the detection chamber of a closed-path instrument. This involves two related issues. First, how does the flow within the tube affect pressure fluctuations between the mouth of the tube and the detection chamber and second, does the presence of the eddy covariance equipment or the creation of a local external flow field caused by pulling the sample into the tube affect or distort the unperturbed ambient atmospheric pressure fluctuations? Each of these questions is examined in turn.

4.3.1 Pressure fluctuations and tube flow

For eddy covariance applications ($f \leq 20$ Hz) the tube acts as a first order filter when the flow is uniform and laminar or nonturbulent (Iberall 1950, Holman 2001). The corresponding complex transfer function, $h_{p'}(\omega)$, and associated first order time constant, $\tau_{p'}$, for the attenuation of pressure fluctuations by uniform laminar tube flow are

$$h_{p'}(\omega) = \frac{1}{1 - j\omega\tau_{p'}} \qquad (7.9)$$

$$\tau_{p'} = \frac{8\mu L_t}{\pi a^4} \frac{V}{\gamma \bar{p}_a} \qquad (7.10)$$

where μ [$\approx 0.18(10^{-4})$ Pa s] is the dynamic viscosity of air, a is the tube radius, V is the volume of the detection chamber, and $\gamma = 1.4$ is the ratio of C_p to C_v for air. (Here C_p and C_v are the specific heats of air at constant pressure and volume.)

The time constant, $\tau_{p'}$, of a system defined by a Licor 6262 with internal pressure, $\bar{p}_a$, of about 96 kPa attached to a tube of length 10 m and inside diameter of 6.35 mm is approximately 0.0004 s, which suggests that for most eddy covariance applications the pressure fluctuations are negligibly attenuated by uniform laminar tube flow. But turbulent tube flow tends to increase $\tau_{p'}$ and the resulting attenuation (Rohmann et al. 1957, Brown et al. 1969). As the flow Reynolds number increases $\tau_{p'}$ increases from a few percent (Rohmann et al. 1957) to maybe an order of magnitude or slightly more (Brown et al. 1969). Even so the theoretical τ'_p should be quite short and pressure attenuation could be fairly small for any closed-path eddy covariance system that employs an unobstructed intake tube.

However, p'_a will be attenuated by flow obstructions in the tube, e.g., filters, insects and, even dust in sufficient amounts, and such obstructions can significantly attenuate pressure fluctuations (Bedard 1977). To further complicate the issue of pressure fluctuations inside the sample chamber, the combined volumes of the tube and detection chamber can act as a resonance cavity (Aydin 1998, Holman 2001), with a resonance frequency $f_n = n\sqrt{3\pi a^2 C_s^2/(4L_t V)}/(2\pi)$, where C_s is the speed of sound and $n = 1, 2, 3, \ldots$, specifies the harmonic frequency. For the present example $f_1 \approx 24.5$ Hz suggesting that these higher frequency pressure fluctuations would be amplified in the detection chamber. Another possible source of high frequency pressure fluctuations is the pump, which for closed-path eddy covariance systems should be downstream of the detection chamber. These pumps typically are diaphragm pumps which operate at 50 or 60 Hz, which, if some design precautions in the tubing connecting the pump and the detection chamber are not taken, could contaminate the detection chamber with 50 or 60 Hz pressure fluctuations. In general, the attenuation due to flow obstructions is likely to dominate any high frequency resonance effects because the spectral power in this high frequency range is too small to be of much concern.

Overall it seems that tube flow employed by closed-path eddy covariance systems is likely to have some impact on the transmission of pressure fluctuations at frequencies below about 10 or 20 Hz. This is due either to the intentional use of filters or the unintentional introduction of tube obstructions resulting from field deployment. Consequently, at frequencies of major concern to eddy covariance flux measurements p'_a amplitudes inside the detection chamber may be reduced relative to the p'_a amplitudes just outside the mouth of the intake tube. But they cannot ever be fully eliminated. Therefore, as with the open-path system, the WPL80 pressure covariance term may also be important to estimates of surface fluxes for closed-path systems during windy, turbulent conditions. However, without some ability to quantify the within-chamber high-frequency p'_a it may not be possible to precisely determine how important these effects could be.

4.3.2 Possible influence of the instrumentation on ambient pressure fluctuations at the tube mouth

The sonic anemometer, the instrumentation mounting structure, the mouth of the tube, and the flow field created by the intake system can interact with the local ambient flow field to create dynamic pressure fluctuations near the mouth of the intake tube. For example, eddies can be shed from the equipment or the mounting boom when the Strouhal number is about 0.2. [The Strouhal number is the nondimensional frequency used to describe vortex shedding. It is defined as fL/U, where f [Hz] is frequency, L [m] is a characteristic length scale of the vortex-shedding object, and U [m s^{-1}] is the speed of the wind impinging upon the object.] Assuming a characteristic length scale of 0.05 to 0.25 m for the eddy covariance equipment and a typical wind speed between 2 to 8 m s^{-1}, then the characteristic eddy shedding frequency could be anywhere between about 2 and 30 Hz. Conceivably, the associated dynamically-induced pressure fluctuations could suppress or enhance any ambient atmospheric static pressure fluctuations that may be present naturally. There are also internal and external tube boundary layers that are created by the ambient flow that will depend on the wind direction and speed (e. g., Kim et al. 2001). However, these effects are likely to be relatively small scale and confined to high frequencies. But, there may also be larger quasi-static pressure fields that are formed by the interaction of the instruments and the wind, which would likewise be a function of wind speed and direction. A full discussion of this issue is beyond the intention of the present study, but it is important that this possibility be mentioned as a research need for closed-path systems. But, in general, the discussions just presented suggest that p'_a within the chamber may or may not reflect the true atmospheric static p'_a, so that estimating $\overline{w'p'_a}$ from ambient measurements, which is exampled in the next section, may or may not provide an accurate estimate of the covariance between w' and p'_a inside the detection chamber.

4.4 Synthesis: Possible consequences for flux estimates

The traditional application of the WPL80 theory to closed-path systems assumes that the temperature covariance term has been eliminated because $T'_a = 0$ within the detection chamber, that the pressure covariance term never contributes because $p'_a/\overline{p_a}$ is negligible, and that any spectral correction to the vapor covariance term includes the tube attenuation effects. This approach basically uses the mass density and water vapor measurements to form covariances and then combines the results to estimate the surface mass flux, $\overline{\rho_d}\overline{w'\omega'_g}$. An alternative to this approach is to convert the measured mass density to mass mixing ratio at the high frequency data rate and then to estimate the surface mass flux by decomposing ω_g into its mean and fluctuating parts and calculating $\overline{w'\omega'_g}$ directly. This section applies the insights developed earlier to these two approaches. For the first approach a numerical example is provided. In the second, the discussion outlines possible discrepancies with the first approach.

4.4.1 Influence of spectral corrections and pressure fluctuations

This subsection estimates the errors in estimates of the surface flux associated with ignoring the WPL80 pressure covariance and overcorrecting the water vapor covariance. Including the tube attenuation as part of the spectral correction to $\mu_v \overline{\omega}_g \overline{w'\rho'_v}$ overcorrects the spectral attenuation by an amount $A_{wv}/A^d_{wv} - 1$. Combining this overestimate with the pressure covariance term yields the following expression for the error, $\Delta(\overline{\rho_d}\overline{w'\omega'_g})$, in the estimate of $\overline{\rho_d}\overline{w'\omega'_g}$ resulting from a misapplication of

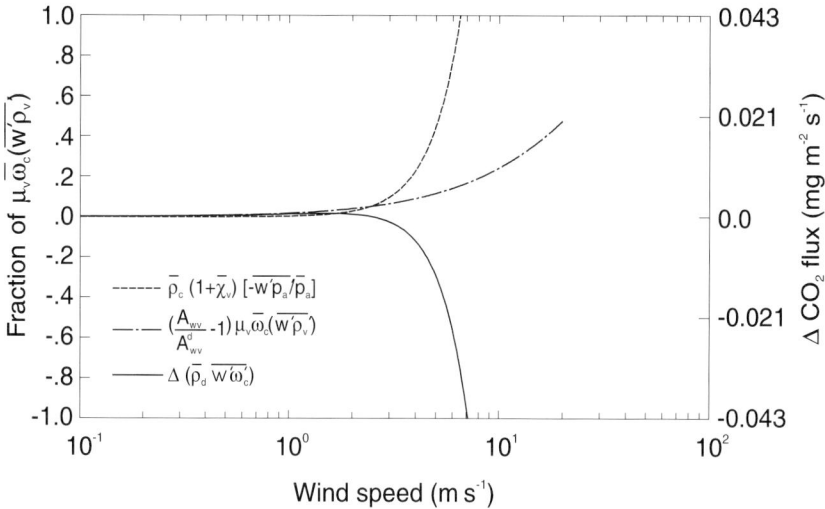

Figure 7.4. Equivalent CO$_2$ flux magnitudes and wind speed dependencies for the various terms of Equation 7.11, showing the consequences of ignoring the WPL80 pressure covariance term and of overcorrecting the vapor covariance term. The combined effect (solid line) indicates that the surface flux for CO$_2$, $\overline{\rho}_d \overline{w'\omega'_c}$, is underestimated (negative quantities). For the purposes of comparison note that 0.02 mg CO$_2$ m^{-2} s^{-1} = 1.72 tC ha^{-1} yr^{-1}.

the WPL80 theory:

$$\Delta(\overline{\rho}_d \overline{w'\omega'_g}) = \frac{A_{wv}}{A_{wv}^d} \mu_v \overline{\omega}_g \overline{w'\rho'_v} - \left\{ \mu_v \overline{\omega}_g \overline{w'\rho'_v} + \overline{\rho}_g (1 + \overline{\chi}_v) \left[-\frac{\overline{w'p'_a}}{\overline{p}_a} \right] \right\} \quad (7.11)$$

This error and its components are evaluated numerically using the same scenario and assumptions listed in section 4.1.3 for Figure 7.3. However, it is more convenient to express the pressure covariance in terms of the wind speed, u. This is done first by noting that $-\overline{w'p'_a} = C\overline{\rho}_a u_*^3$, where $C \approx 2$ for neutral atmospheric conditions (Wilczak et al. 1999, Massman and Lee 2002), and second by assuming that $u_* = Bu$, where $B \approx 0.2$ for forested canopies and $B \approx 0.1$ is more appropriate for agricultural crops. Note that this relationship between u and u_* does not necessarily apply universally. It is useful here for numerical purposes only and should not be taken as indicative of any particular site, where it will depend upon the measurement height, the atmospheric stability, the canopy roughness length, etc. For estimating $-\overline{w'p'_a}$ at any given eddy covariance site the relationship $-\overline{w'p'_a} = C\overline{\rho}_a u_*^3$ should use u_* values

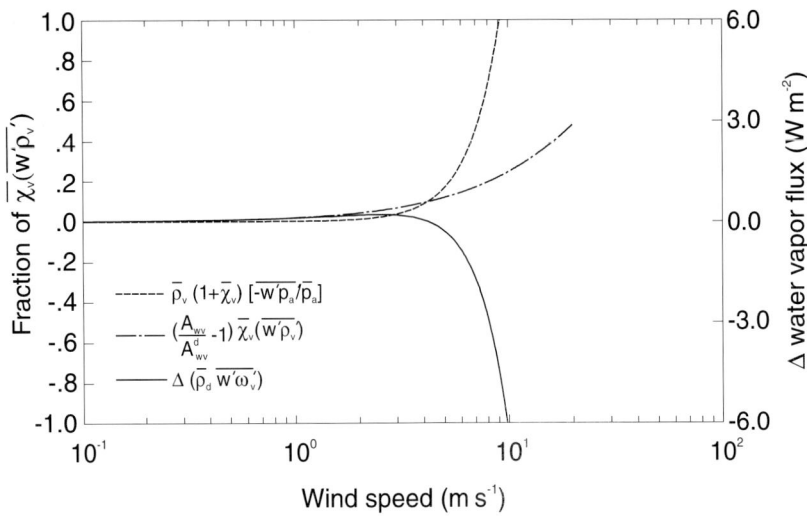

Figure 7.5. Equivalent water vapor flux magnitudes and wind speed dependencies for various terms of Equation 7.11, showing the consequences of ignoring the WPL80 pressure covariance term and of overcorrecting the vapor covariance term. The combined effect (solid line) is that the true surface flux for water vapor, $\overline{\rho}_d \overline{w'\omega'_v}$, is underestimated (negative quantities). For this example the water vapor flux is assumed to be about 300 W m^{-2}.

measured with the sonic anemometer rather than estimating u_* from the wind speed. However, the multiplier C increases as the atmosphere becomes more unstable (Wilczak et al. 1999).

The numerical evaluation of Equation 7.11 is performed for both CO_2 and water vapor and assumes: $\mu_v = 0.622$, $\overline{\omega}_c = 0.57$ mg g^{-1}, $\overline{w'\rho'_v} = 0.12$ g m^{-2} s$^{-1} \approx 300$ W m^{-2}, $\overline{\rho}_c = 730$ mg m^{-3}, $\overline{\chi}_v = 0.02$, $\overline{\rho}_a = 1.28$ kg m^{-3}, $\overline{p}_a = 100$ kPa, and $\overline{\rho}_v = 15$ g m^{-3}. The ratio A_{wv}/A^d_{wv} is computed following the integral approach for estimating correction factors (see Equation 3 of Massman 2000). For A^d_{wv} is it assumed that the first order response time of the closed-path system (tube + Licor 6262) was determined empirically to be 0.3 s. Thus the equivalent response time includes the tube attenuation as well as the 6262's signal processing software and its volume averaging effects. For A_{wv} only the third order Bessel filter and the volume averaging effects, both discussed earlier, are used.

The results for CO_2 are shown in Figure 7.4 and water vapor in Figure 7.5. Each figure includes $\Delta(\overline{\rho}_d \overline{w'\omega'_g})$, $[\frac{A_{wv}}{A^d_{wv}} - 1]\mu_v \overline{\omega}_g \overline{w'\rho'_v}$, and the pressure covariance term, $\overline{\rho}_g (1+\overline{\chi}_v)[-\overline{w'p'_a}/\overline{p}_a]$. Each expression is evaluated as a fraction of $\mu_v \bar{\omega}_g \overline{w'\rho'_v}$ (left axis) and an absolute amount (right axis) for the case $B = 0.2$.

To aid in the interpretation of Figure 7.4 it should be noted that 0.02 mg CO_2 m^{-2} s^{-1} = 1.72 tC ha^{-1} yr^{-1}, suggesting the potential for significant biases to the annual carbon budget. Nevertheless, it is also important to note that this figure is valid for the instantaneous half-hourly or hourly fluxes and does not directly relate to the annual flux. The consequences of these errors to the annual carbon sum can be estimated by extending the present example to include an annual growing cycle.

Assume that the daily maximum in the water vapor flux is 300 W m^{-2}, and that a typical diel cycle follows a sine wave that is 12 hours long. Further assume that the growing season is also sinusoidal and 6 months long. Averaging the instantaneous bias introduced solely be the overestimation of the water vapor term over the daily and seasonal cycles reduces the instantaneous bias by about a factor of about 10 (or more precisely by a factor of π^2). So that the corresponding annual carbon sum would be biased too high by about 0.04 to 0.16 tC ha^{-1} yr^{-1} (depending on the half-hourly or hourly mean wind speed) solely as a result of the misapplication of the spectral correction to the WPL water vapor term. The potential bias introduced by ignoring the pressure covariance could be larger and in the opposite sense, but does depends even more critically on the aerodynamic nature of the surface and the frequency of high wind during the growing season. Because these two terms tend to compensate a conservative (and relatively low-wind speed) estimate of the uncertainty introduced into the annual carbon balance by these two WPL terms alone can reasonably be assumed to be about ± 0.1 tC ha^{-1} yr^{-1}. However, this uncertainty could easily become a much larger bias at higher wind speeds and rougher surfaces.

Consequently, even small biases resulting from these discrepancies can lead to potentially significant biases in the annual carbon budget estimated by eddy covariance. In general, these CO_2 results suggest that ignoring the pressure covariance term introduces a larger bias into the estimate of the surface CO_2 flux, $\bar{\rho}_d \overline{w'\omega'_g}$, than overcorrecting the water vapor covariance term. But, the overcorrected water vapor covariance does partially compensate for the lack of the pressure covariance term. Figure 7.5 suggests that the consequences to the surface water vapor flux are similar, but less significant than for the CO_2 surface flux. Figure 7.5 also indicates that ignoring the pressure covariance term can cause the lack of closure (underestimation) of the surface energy balance to worsen as wind speed increases.

For the case $B = 0.1$ the results (not shown) were similar to those shown in these last two figures, except that the pressure covariance term, although still significant to the surface flux estimates, was reduced relative to the $B = 0.2$ scenario. Finally, it is important to reiterate that all results presented in this section are intended as plausible examples only. They are useful for indicating general features and general consequences. But the specific numerical results do not necessarily apply universally, because each eddy covariance site is likely to have different sensor deployment, potentially different sensor time constants, different measurement heights, and different data processing algorithms. The same caveat is true for the next section.

4.4.2 High frequency conversion of mass density to mixing ratio

This section examines the consequences of estimating the surface flux by converting the high frequency data, point-by-point to $w_g = \rho_g/\rho_d$, then decomposing it to $\overline{w}_g + w'_g$, and finally using w'_g to form $\overline{\rho_d w' w'_g}$. For this case the WPL80 theory still applies so that $w'_g = \rho'_g + \overline{\rho}_g(1+\overline{\chi}_v)[-p'_a/\overline{p}_a] + \mu_v \overline{w}_g \rho'_v$. But no single instrument measures w_g directly, rather it can only be determined by combining data (or data streams) from more than one instrument. Consequently, it is $\rho'_g + \overline{\rho}_g(1+\overline{\chi}_v)[-p'_a/\overline{p}_a] + \mu_v \overline{w}_g \rho'_v$ that is being measured, not w'_g. Therefore, when forming the covariances it is still appropriate to be concerned with how and with what instruments are the quantities ρ_g, ρ'_g, p'_a, and ρ'_v being measured or calculated. This issue must be addressed if spectral corrections are to be applied appropriately and if the pressure fluctuations need to be included. In general, Equation 7.2 still applies when estimating $\rho_d \overline{w' w'_g}$.

Consider the following, and final, example. Assume that mean pressure inside the tube is known, but measured with a relatively slow response sensor so that p'_a cannot be measured and the pressure covariance term is thereby implicitly ignored. Further assume that all other conditions and parameter values are the same as those already provided in the previous example except that the response time of the CO_2 sampling system has been found empirically to be 0.3 s and for water vapor the response time was found to be 0.5 s. In this case applying the correction factor associated with $\overline{w' \rho'_c}$ to $(\overline{\rho}_d \overline{w' w'_c})_m$ would yield the same result as shown in Figure 7.4. This approach would properly correct the measured CO_2 density covariance, $A^d_{wc} \overline{w' \rho'_c}$, but would again overestimate the vapor covariance term exactly as shown in Figure 7.4. For water vapor the results are similar to those shown in Figure 7.5, except that the overestimation factor, A_{wv}/A_{wv}^d, is now about 50% greater, a consequence of using a response time of 0.5 s rather than 0.3 s.

These last two examples indicate that the application of WPL80 terms and spectral corrections to closed-path eddy covariance systems do not commute and that the preferred approach must be first to apply appropriate spectral corrections to each of the terms in Equation 7.2, then add them together to form the surface flux. If this is not done the use of an active trace gas sampling system will virtually guarantee that any estimate of the surface flux will be biased because the amount of attenuation of the density covariance term, A_{wg}^d, is likely to exceed the amount of attenuation of the WPL80 vapor covariance term, A_{wv}. In other words, it is very important to estimates of surface flux not to confuse combining data streams from different instruments, which is a mathematical operation, with making a direct measurement of ω_g with a single instrument.

4.5 Low frequency temperature fluctuations

Contamination of the closed-path sampling chamber by low frequency temperature fluctuations can occur under certain conditions (Leuning 2003, personal communication). This is contrary to what is normally assumed for high-frequency temperature fluctuations and for closed-path systems in general. Such contamination is most likely to occur when low frequency convective motions carry some portion of the heat flux (e.g., Sakai *et al.* 2001, Finnigan *et al.* 2003, Chapter 5). However, the intake tube also plays a role in this type of contamination. If the intake tube is isothermal during any flux averaging period it will act as a low pass filter in regards to these low frequency temperature fluctuations. However, if there are longitudinal or temporal variations in the temperature of the tube walls during a flux averaging period, then low frequency temperature contamination of the sample is much more complex than can be described by low pass filtering. In this case even the sign of the leakage term is uncertain. In general, the amount of the low frequency T_a' leakage is determined by the nature of the external heat exchange between the tube and its ambient environment, the thermal and radiational properties of the tube material, the tube flow rate, and the length of the tube. Of course, no amount of temperature or energy exchange between the sample and tube walls is of any importance unless it correlates with the fluctuations in the vertical wind speed.

To further complicate this matter, low frequency T_a' contamination of closed-path systems is likely to be site specific. This is because low frequency atmospheric motions are likely to be tied to the nature of the aerodynamics and the heating of the underlying surface and to the design, implementation, and material of the intake tube. Nevertheless, if a little as 1% of the total (300 W m^{-2}) heat flux leaks through to the detection chamber the effect on the instantaneous CO_2 flux, discussed in the example in the previous section, is about 0.01 mg CO_2 m^{-2} s^{-1}. (This calculation was not included in the figures of the previous section because of the large uncertainty involved in estimating the amount of the leakage.) If as much as 5% of the heat flux leaks into the detection chamber, then it is likely to dominate all other effects and result in a very large underestimation of the closed-path eddy covariance annual carbon balance. Clearly, this issue has the potential to be quite serious for closed-path eddy covariance CO_2 fluxes and as such needs further research.

5 Summary and Conclusions

Open- and closed-path CO_2 and water vapor eddy covariance systems are similar in their use of infrared gas analyzers to measure trace gas fluctuations. But, they are different in their handling of the air being sampled. These differences are crucial when applying spectral corrections and the WPL80 terms for flux estimation. Open-path systems are purely passive, i. e., they do not physically alter the sample. Whereas closed-path systems combine aspects of both active and passive sampling with the intake tube acting as the active portion. It is the active portion of the system that physically alters the sample by eliminating the temperature fluctuations and attenuating the water vapor and CO_2 fluctuations through a combination of diffusional smoothing and interaction with the tube walls.

The spectral corrections associated with passive sampling describe instrument or data processing compromises and they apply to all covariances (including the WPL80 terms) and to either an open- or closed-path system. However, these corrections are specific to a particular instrument and data processing system and they are not necessarily the same for any of the covariances: $\overline{w'T'_a}$, $\overline{w'p'_a}$, $\overline{w'\rho'_c}$, or $\overline{w'\rho'_v}$. Spectral corrections associated with active sampling describe sample-handling compromises and they apply only to the density covariance term, $\overline{w'\rho'_g}$, not to the (closed-path-associated) WPL80 vapor or pressure covariance terms. This is a consequence of the fact that the WPL80 terms characterize the environment in which the trace gas measurements are made. In the case of the open-path the WPL80 covariance terms can be interpreted as fluxes (after spectral correction). In the case of the closed-path the WPL80 covariance terms lose their interpretation as fluxes, because fluctuations in temperature, pressure, and water vapor of the air being sampled by the detection chamber have been physically altered by the tube and the tube flow.

This study has attempted to provide a template for the application of spectral corrections to the WPL80 terms and the estimation of fluxes by reexamining the original WPL80 theory from the perspective of the instrumentation and its supporting technology. The major conclusions are

- With current technology the application of spectral corrections and the WPL80 terms do not commute and spectral corrections should be made to all covariances first before summing the WPL80 terms to estimate surface fluxes.

- High frequency point-by-point conversions from mass density to mixing ratio is not the preferred method for estimating fluxes by eddy covariance.

- For closed-path systems the spectral corrections for the WPL80 covariance terms and the density covariance term, $\overline{w'\rho'_g}$, are not the same.

- For some atmospheric conditions the WPL80 pressure covariance term, which is usually ignored, can be important for closed-path estimates of both the CO_2 flux and the surface energy balance. This is because pressure fluctuations are usually not measured within the detection chamber so their significance cannot be quantified very precisely. They may be enhanced over the ambient external flow and they not suffer significant attenuation with tube flow.

- Using the same spectral corrections for the density covariance term, $\overline{w'\rho'_g}$, and the WPL80 water vapor covariance term can introduce biases into the annual estimates of the carbon balance, as can ignoring the WPL80 pressure covariance term. However, possible contamination of the closed-path detection chamber by low frequency temperature fluctuations may be the largest source of bias in annual carbon balances measured with closed-path systems.

6 Acknowledgment

This work evolved from the August 2002 AmeriFlux workshop on the standardization of flux analysis and diagnostics. The author extends his thanks to D. Billesbach, S. Miller, B. Amiro, X. Lee, R. Leuning, S. Wofsy, and T. A. Black for their discussions, comments, and insights and to Licor Inc. and to R. Eckles and D. Anderson in particular for their assistance with the third order Bessel filter. Thanks also go to A. Bedard and J. Wilczak for many helpful discussions concerning the nature and measurement of pressure fluctuations and to R. Kelly for discussions concerning some observed pressure fluctuations inside the Licor 6262 that is flown on the King Air.

7 References

Andreas, E. L.: 1981, 'The effects of volume averaging on spectra measured with a Lyman-Alpha hygrometer', *J. Appl. Meteorol.* **20**, 467–475.

Aydin, I.: 1998, 'Evaluation of fluctuating pressure measured with connection tubes', *J. Hydraulic Eng.* **124**, 413–418.

Bedard, A. J.: 1977, 'The d-c pressure summator: theoretical operation, experimental tests and possible practical uses', *Fluidics Quarterly*, **1**, 26-51.

Brown, F. T., Margolis, D. L., Shah, R. P.: 1969, 'Small-amplitude frequency behavior of fluid lines with turbulent flow', *Transactions ASME*, **91**, 678–693.

Finnigan, J. J., Clement, R., Mahli, Y., Leuning, R., Cleugh, H. A.: 2003, 'A re-evaluation of long-term flux measurement techniques: Part 1. Averaging and coordinate rotation', *Bound.-Layer Meteorol.* **107**, 1-48.

Frost, S. R.: 1981, *Temperature Dispersion in Turbulent Pipe Flow*, Ph.D. Thesis, University of Cambridge, Cambridge.

Fuehrer, P. L., Friehe, C. A.: 2002, 'Flux corrections revisited', *Bound.-Layer Meteorol.* **102**, 415–457.

Gurvich, A. S.: 1962, 'The pulsation spectra of the vertical component of the wind velocity and their relations to micrometeorological conditions', *Izvestiya Atmospheric and Oceanic Physics*, **4**, 101–136.

Hicks, B. B.: 1972, 'Propellor anemometers as sensors of atmospheric turbulence', *Bound.-Layer Meteorol.* **3**, 214–228.

Holman, J. P.: 2001, *Experimental Methods for Engineers.* 7th Edition. McGraw-Hill, Boston, MA, 698 pp.

Horst, T. W.: 1997, 'A simple formula for attenuation of eddy fluxes measured with first-order-response scalar sensors', *Bound.-Layer Meteorol.* **82**, 219–233.

Iberall, A. S.: 1950, 'Attenuation of oscillatory pressures in instrument lines', *J. Research National Bureau Standards*, **45**, 85–108.

Kaimal, J. C., Wyngaard, J. C., Izumi, Y., Cotè, O. R.: 1972, 'Spectral characteristics of surface-layer turbulence', *Quart. J. R. Meteorol. Soc.* **98**, 563–589.

Kim, Y., Engeda, A., Aungier, R., Direnzi: 2001, 'The influence of inlet flow distortion on the performance of a centrifugal compressor and the development of an improved inlet using numerical simulations', *Proceedings Institute Mechanical Engineers*, **215, Part A**, 323-338.

Lee, X., Black, A., Novak, M. D.: 1994, 'Comparison of flux measurements with open- and closed-path gas analyzers above an agricultural field and a forest floor', *Bound.-Layer Meteorol.* **67**, 195–202.

Leuning, R., Denmead, O. T., Lang, A. R. G., Ohtaki, E.: 1982, 'Effects of heat and water vapor transport on eddy covariance measurements of CO_2 fluxes', *Bound.-Layer Meteorol.* **23**, 209–222.

Leuning, R., Judd, M. J.: 1996, 'The relative merits of open- and closed-path analysers for measurement of eddy fluxes', *Global Change Biology*, **2**, 241–253.

Leuning, R., King, K. M.: 1992, 'Comparison of eddy-covariance measurements of CO_2 fluxes by open- and closed-path CO_2 analysers', *Bound.-Layer Meteorol.* **59**, 297–311.

Leuning, R., Moncrieff, J.: 1990, 'Eddy-covariance CO_2 flux measurements using open- and closed-path CO_2 analysers: Corrections for analyser water vapor sensitivity and damping of fluctuations in air sampling tubes', *Bound.-Layer Meteorol.* **53**, 63–76.

Massman, W. J.: 1991, 'The attenuation of concentration fluctuations in turbulent flow through a tube', *J. Geophys. Res.* **96**, 15,269–15,273.

Massman, W. J.: 2000, 'A simple method for estimating frequency response corrections for eddy covariance systems', *Agric. For. Meteorol.* **104**, 185–198.

Massman, W. J.: 2001, 'Reply to comment by Rannik on "A simple method for estimating frequency response corrections for eddy covariance systems"', *Agric. For. Meteorol.* **107**, 247–251.

Massman, W. J., Lee, X.: 2002, 'Eddy covariance flux corrections and uncertainties in long-term studies of carbon and energy exchanges', *Agric. For. Meteorol.* **113**, 121–144.

Moore, C. J.: 1986, 'Frequency response corrections for eddy correlation systems', *Bound.-Layer Meteorol.* **37**, 17–35.

Paw U, K. T., Baldocchi, D. D., Meyers, T. P., Wilson, K. B.: 2000, 'Correction of eddy-covariance measurements incorporating both advective effects and density fluxes', *Bound.-Layer Meteorol.* **97**, 487–511.

Rannik, Ü.: 2001, 'A comment on the paper by W.J. Massman "A simple method for estimating frequency response corrections for eddy covariance systems"', *Agric. For. Meteorol.* **107**, 241–245.

Rannik, Ü., Vesala, T., Keskinen, R.: 1997, 'On the damping of temperature fluctuations in a circular tube relevant to the eddy covariance measurement technique', *J. Geophys. Res.* **102**, 12,789–12,794.

Rohmann, C. P., Grogan, E. C.: 1957, 'On the dynamics of pneumatic transmission lines', *Transactions ASME*, **79**, 853–874.

Sakai, R. K., Fitzjarrald, D. R., Moore, K. E.: 2001, 'Importance of low-frequency contributions to eddy fluxes observed over rough surfaces', *J. Appl. Meteorol.* **40**, 2178-2192.

Silverman, B. A.: 1968, 'The effect of spatial averaging on spectrum estimation', *J. Appl. Meteorol.* **7**, 168–172.

Suyker, A. E., Verma, S. B.: 1993, 'Eddy covariance measurement of CO_2 flux using a closed-path sensor: theory and field tests against an open-path sensor', *Bound.-Layer Meteorol.* **64**, 391–407.

Webb, E. K., Pearman, G. I., Leuning, R.: 1980, 'Correction of flux measurements for density effects due to heat and water vapor transfer', *Quart. J. R. Meteorol. Soc.* **106**, 85–106.

Wilczak, J. M., Edson, J. B., Høgstrup, J., Hara, T.: 1999, 'The budget of turbulent kinetic energy in the marine atmospheric surface layer', In: Geernaert, GL (Ed.), *Air-Sea Exchange: Physics, Chemistry, and Dynamics*, Kluwer Academic Publishers, Dordrecht, The Netherlands, 153–173.

Chapter 8

STATIONARITY, HOMOGENEITY, AND ERGODICITY IN CANOPY TURBULENCE

Gabriel Katul, Daniela Cava, Davide Poggi, John Albertson, Larry Mahrt

gaby@duke.edu

Abstract One of the defining syndromes of turbulence is nonlinear stochasticity. This view of turbulence motivated the development of statistical mechanics theories that have served to connect the basic Navier-Stokes (NS) equations of motion to the statistical results of numerous field experiments. In general, the proper averaging operator for stochastic processes is ensemble averaging. Given the transient nature of flow boundary conditions in natural systems, field experiments are typically unable to capture a suitable ensemble, in a strict sense. Instead, field experiments typically focus on time averaged statistics. Stationarity and ergodicity are two central concepts (required conditions) used to link field measurements and the NS equations or field measurements to "boundary conditions" at the land-atmosphere interface. In this Chapter, we present an elementary review of these two concepts for the atmospheric surface layer (ASL) and canopy sublayer (CSL) and proceed to show why the stable CSL tends to violate both conditions. A weaker form of these two conditions may be applicable to CSL flows that are only moderately stably stratified. Practical implications for nighttime CO_2 flux corrections are also discussed.

1 Introduction

It has been argued that the Navier-Stokes (NS) equations provide a mathematical basis of turbulence (Monin and Yaglom 1971). For high Reynolds number flows, such as atmospheric boundary layer flows, these equations exhibit extreme sensitivity to initial and boundary conditions. This sensitivity has important consequences for how NS averaging must be performed and under what conditions samples taken over sequential periods of time, such as in field experiments, may be considered to represent an ensemble of the process for a given set of initial and boundary conditions. In this Chapter we explore these issues and discuss their connection to the nighttime CO_2 flux correction problem. This Chapter is not intended as a formal review of stationarity, homogeneity, and ergodicity of NS; rather, it is intended to show that barriers to our progress in interpreting longterm eddy-covariance fluxes, particularly nighttime CO_2 fluxes, arise because of non-stationarity and lack of ergodicity.

The underlying importance of this issue is highlighted by the following scenario: Instruments on a flux tower are used to compute time averaged statistics (e. g. covariances) and these statistics are taken to be direct representations of the influence of the land surface boundary conditions on the flow. Implicit in this assumption is a belief that if this experiment were repeated with the same boundary conditions, then the same statistics would be observed. However, if the necessary conditions for ergodicity were not satisfied, then the above belief would be invalid.

The sensitivity of turbulence to initial conditions has been examined both numerically and through laboratory measurements of velocity. Here, velocity measurements are assumed to represent solutions to NS (Lesieur 1990, Frisch 1995). When wind tunnel experiments are repeated (N times) without changing the boundary conditions, the details of the velocity time series differ across the N realizations (Lesieur 1990). Hence, turbulence is often characterized as a stochastic phenomenon (Tennekes and Lumley 1972). On the other hand, closer scrutiny of the same N experiments would suggest that the statistical properties of the velocity are reproducible (Frisch 1995). If each of these N experiments is a solution to the NS equations, then a logical question is "how can randomness arise across the experiments if the experimental configuration and steps are unchanged?"

To address this question, we use a toy model known as the *Poor-Man's Navier-Stokes Equation* (Frisch 1995). In this model, the NS equations are replaced with a logistic map model possessing the following attributes: 1) a linear difference term to replace the local acceleration, 2) a second-order nonlinear term to replace the advective acceleration and the pressure gradient, 3) a linear term to replace the viscosity effects, and 4) a constant to replace body forces. That is

$$\overbrace{u(t+1) - u(t)}^{1} = \overbrace{-2u(t)^2}^{2} \overbrace{-u(t)}^{3} \overbrace{+1}^{4}$$

$$\overbrace{\frac{\partial u_i}{\partial t}}^{1} = -\overbrace{\left[u_j \frac{\partial u_i}{\partial x_i} + \frac{\partial p}{\partial x_i}\right]}^{2} + \overbrace{\nu \frac{\partial^2 u_i}{\partial x_j \partial x_j}}^{3} \overbrace{+ f_i}^{4} \qquad (8.1)$$

Stationarity, Homogeneity, and Ergodicity 163

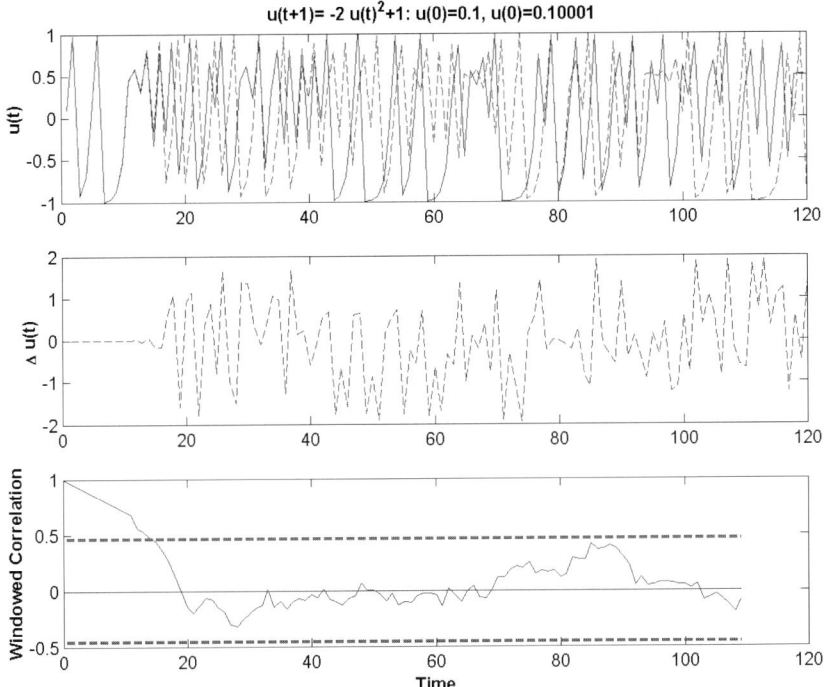

Figure 8.1. Top: solutions to the logistic equation for $u(0) = 0.1$ and $u(0) = 0.10001$ as a function of time units; Middle: the variation of the solution difference (Δu) with time units. Note that $|\Delta u|$ can be larger than u; Bottom: the variation of the windowed correlation coefficient for the two solutions in the top panel with the time origin. The dotted lines are the 99% probability levels for which the two time series are independent.

where t is time, x_i are Cartesian spatial coordinates ($x_1 = x$, $x_2 = y$, and $x_3 = z$) with x_1 being the longitudinal direction, x_2 being the lateral direction, and x_3 being the vertical direction, u_i are the velocity components along x_i, p is pressure, ν is kinematic viscosity, and f_i is a body force. The above logistic equation is de-coupled from any turbulence physics for numerous reasons (e. g. it has no spatial structure); however, it generates time series with similar degrees of nonlinearity as the NS equation and, therefore, use of this model allows us to explore the underlying need for a probabilistic description of turbulence. For example, in Figure 8.1, the solution to the logistic equation for two initial conditions: $u(0) = 0.1$ and $u(0) = 0.10001$ are shown (hence, the two cases start with a difference in initial conditions of one part in 10,000 —

very likely immeasurable). The difference between these two solutions is plotted as a function of time in the middle panel of Figure 8.1. As expected for a system with sensitive dependence to initial conditions, the two solutions diverge after a short period and the temporal pattern details become drastically different. This point is made clear by the temporal evolution of the windowed correlation coefficient between the two time series, plotted in Figure 8.1. Note that the realizations are near perfectly correlation initially and become uncorrelated after a short duration. In fact, the difference between the two solutions is as large as the magnitude of u.

What this example illustrates is how a nonlinear term, like that present in the NS equations, leads variables such as turbulent velocities (and scalar concentrations) to be random variables, thus dictating that the proper averaging of the governing equations must be conducted via ensemble averaging. However, tower-based field experiments measure temporal averages (e. g. covariances between vertical velocity and concentration fluctuations). An obvious fundamental question then is under what conditions do temporal averages converge to ensemble averages? Stated differently, under what conditions is the ergodic hypothesis valid? We will explore this question for both ASL and CSL turbulence, and proceed to show how the stable CSL can drastically violate assumptions necessary for ergodicity. We limit our discussion to uniform flat terrain though planar non-uniformity in canopy density as well as topography are equally critical to describing the CSL (Finnigan et al. 1990, Raupach and Finnigan 1997, Baldocchi et al. 2000, Finnigan 2000, Finnigan et al. 2003). Practical "fixes" such as thresholds based on the friction velocity (u_*) to correct nighttime CO_2 fluxes are shown to be a reasonable starting point, but progress on stable CSL flows requires development of new statistical mechanics methodologies not anchored to the ergodic hypothesis. The necessary conditions for stationarity and ergodicity are briefly described next.

2 Stationarity, Homogeneity, and the Ergodic Hypothesis

Simply stated, a random variable is said to be stationary if all its statistical moments (including joint statistics) are independent of time. To conceptually illustrate stationarity, consider the hypothetical experimental setup in Figure 8.2 in which the longitudinal velocity time series is measured above a cylinder. For the same probe and cylinder configuration in Figure 8.2, repeat the experiment N times (only 3 are shown

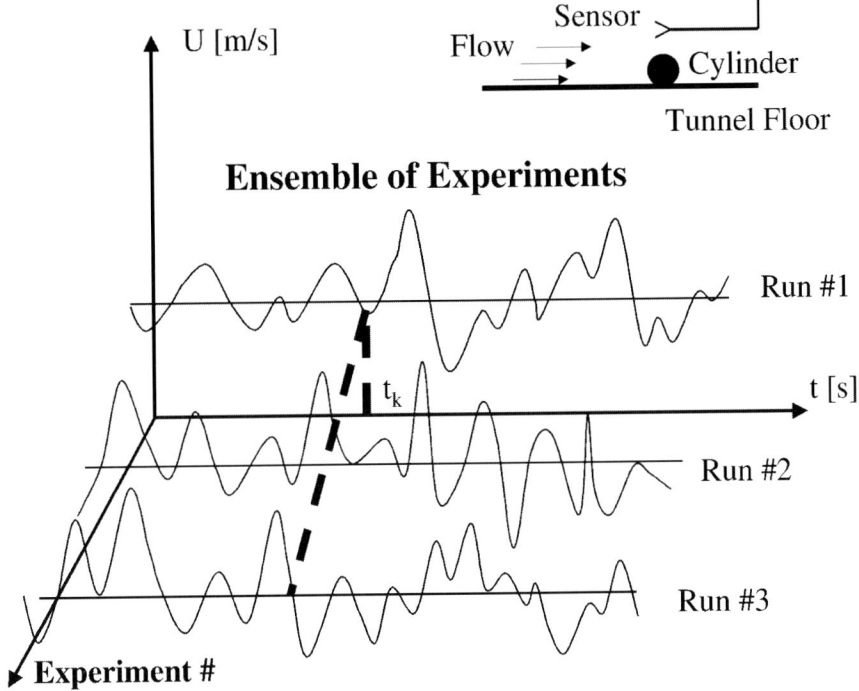

Figure 8.2. Construction of the ensemble of experiments for the velocity time series above the cylinder. Each run is for the same experimental setup.

in Figure 8.2) so as to construct a large ensemble of velocity time series runs, with each experimental run sampling M velocity values.

From this ensemble of experiments the velocity is said to be strictly stationary if at each time instant (with $t = 0$ at the start of each run), the ensemble probability density function from all those experiments is independent of time. That is, at each time t, the statistical moments (including joint moments) are computed across the ensemble of runs; these moments are then evaluated for time dependence.

For the experiments in Figure 8.2, the ensemble average at time t_k for the rth longitudinal velocity moment is given by

$$<u^r(t_k)> = \frac{1}{N}\sum_{j=1}^{N}[^j u(t_k)]^r \qquad (8.2)$$

where $^j u(t_k)$ is the measured velocity at time t_k from experiment j ($j = 1, 2...N$).

Also, the ensemble joint moment can be computed as

$$<u^r(t_k)u^q(t_s)> = \frac{1}{N}\sum_{j=1}^{N}\{[^j u(t_k)]^r[^j u(t_s)]^q\} = g_{r,q}(t_k,t_s) \quad (8.3)$$

where t_k and t_s are any arbitrary time points. In theory, Equations 8.2 and 8.3 must be evaluated in the limit when $N \to \infty$; in practice, N must be finite but sufficiently large.

For a strictly stationary process

$$<u^r(t_1)> = <u^r(t_2)> = \ldots = <u^r(t_k)>; k = 1, \ldots M$$
$$<u^r(t_k)u^q(t_s)> = g_{r,q}(t_k - t_s) = g_{r,q}(\tau) \quad (8.4)$$

Equation 8.4 states that all ensemble statistics are independent of time, and vary only with the time lag τ. In practice, it is common to discuss stationarity in a "weak sense" in which only lower-order moments are considered. For a weakly stationary process,

$$<u(t_1)> = <u(t_2)> = \ldots = <u(t_k)>; k = 1, \ldots M$$
$$<u(t_1)^2> = <u(t_2)^2> = \ldots = <u(t_k)^2>; k = 1, \ldots M$$
$$<u(t_k)u(t_s)> = g_{1,1}(\tau); k = 1, \ldots M \quad (8.5)$$

Furthermore, a process is said to be ergodic if it is stationary and if $g_{r,q}(\tau) \to 0$ as $\tau \to \infty$.

It can be shown that for an ergodic process (Stanisic 1985),

$$<u(t_1)^r> = \ldots = <u(t_N)^r> = \ldots = \overline{^1 u^r} = \ldots = \overline{^M u^r}$$
$$<u(t_k)^r u(t_k + \tau)^q> = \overline{[^j u(t)]^r [^j u(t+\tau)]^q} \quad (8.6)$$

for any j, where

$$\overline{^j u^r} = \frac{1}{M}\sum_{l=1}^{M}[^j u(t_l)]^r \quad (8.7)$$

That is, for the setup in Figure 8.2, the ergodic hypothesis states that ensemble statistics (including joint statistics) at any instant in time are identical to temporal statistics from any realization or experiment. Again, analogous to stationarity, ergodicity can be considered in a "weak sense" by focusing on first and second moments (and covariances).

If the time domain is simply replaced with a spatial direction (e. g. x_1), then stationarity is simply replaced by homogeneity along this direction. We consider next how such concepts apply to ASL and CSL flows, respectively.

3 Stationarity, Homogeneity, and Ergodicity in Atmospheric Surface Layer Flows

Conducting "repeatable experiments" in the ASL is difficult because replicating an experiment for the same meteorological and hydrological conditions is next to impossible (Monin and Yaglom 1971). For this reason, Monin and Yaglom (1971) suggested that experiments conducted under "similar" *mean* meteorological conditions be used to define ensembles analogous to Figure 8.2. The concepts of stationarity and ergodicity can be tested for such ensembles of experiments. While testing the validity of such an idealization has not been undertaken to date, recent advances in LIDAR (Light Detection and Ranging) measurements offer a promising first step for direct evaluation of such hypotheses for ASL flows. We note that *local* homogeneity (or stationarity) and isotropy at fine scales (within the so-called inertial subrange) have been extensively studied (Monin and Yaglom 1971, Kaimal and Finnigan 1994, Frisch 1995, Finnigan 2000, Katul et al. 2001, Giostra et al. 2002) and are beyond the scope of this work.

As an example, we show water vapor concentration (q) measurements collected at $z = 3$ m above an irrigated bare soil surface at the Campbell Tract facility in Davis, California. The scanning, solar blind water Raman LIDAR used here was built at Los Alamos National Laboratory and is described elsewhere (Eichinger et al. 1994). In these experiments, the LIDAR sampled with a spatial resolution of 1.5 m and a temporal resolution of 1.0 s for about 9.5 minutes. The spatial range for which the signal to noise ratio exceeds 10 is shown in Figure 8.3, which is roughly 50 m from the LIDAR. The LIDAR experiment can be viewed in one of the following two setups, both constructed under similar mean meteorological conditions

- An ensemble of 35 towers (arrayed in space) sampling the temporal structure of turbulence on a 1 s time step.

- An ensemble of 543 towers (arrayed in time) sampling the spatial structure of turbulence on a 1.5 m spatial resolution.

From Figure 8.3, if scenario 1 is adopted, then it is possible to test whether

$$< q(x_1) >=< q(x_2) >= ... =< q(x_k) >; k = 1,...35$$
$$< q(x_1)^2 >=< q(x_2)^2 >= ... =< q(x_k)^2 >; k = 1,..35 \quad (8.8)$$

That is, this test serves to assess whether the ASL is (weakly) homogeneous along the LIDAR beam direction. Also, if scenario 2 is adopted,

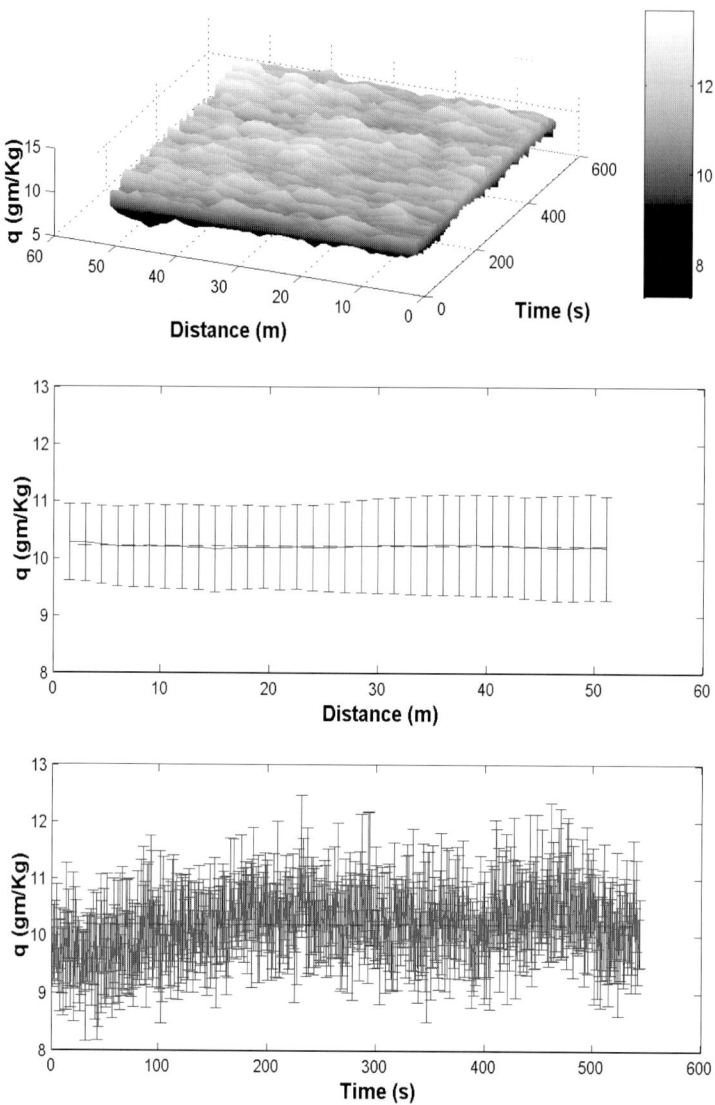

Figure 8.3. Top: The space-time variation of the LIDAR measured water vapor concentration (q) at $z = 3$ m above a uniform irrigated bare soil field; Middle: The variation of the time-averaged q at each spatial location obtained from the above panel; Bottom: The variation of the spatially averaged q at each instant in time from the top panel. Vertical error bars are one standard deviation.

then it is possible to test whether

$$< q(x_1) > = < q(x_2) > = ... = < q(x_k) >; k = 1, ...543$$
$$< q(x_1)^2 > = < q(x_2)^2 > = ... = < q(x_k)^2 >; k = 1, ..543 \quad (8.9)$$

Note that in Equations 8.8 and 8.9 we are taking time points and space points, respectively, as proxies for realizations in an ensemble averaging. The results of the homogeneity and stationarity tests are shown in Figure 8.3, showing means that are remarkably stable in both time and space. We compare whether either of these two proxy "ensemble" averages are statistically different from the overall space-time average of the entire q field ($= 10.2 \, \text{g}\,\text{kg}^{-1}$) shown in Figure 8.3 top panel (whose value is shown as dotted line in Figure 8.3 middle and bottom panels). Using the standard student t-test, we tested whether these two ensemble averages are statistically different from $10.2 \, \text{g}\,\text{kg}^{-1}$ and found that the assumptions of stationarity and homogeneity can not be rejected at the 95% confidence level.

Figure 8.4 shows the ensemble autocorrelation function (of ensemble scenario 1) and demonstrates that 1) the spatial variability in autocorrelation (bars) is small compared to the 99% confidence level in statistically significant correlation (range between dashed lines), and 2) the ensemble autocorrelation function decays rapidly with increasing time lag. The 99% confidence limits are computed using (Anderson 1976).

$$r_\tau(99\%) = \frac{-1 \pm 2.326\sqrt{M - \tau - 1}}{M - \tau}$$

where r_τ are the 99% probability levels for which the time series is independent, and M is the sample size.

In short, from this limited LIDAR experiment, it appears that "necessary" conditions for the ergodic hypothesis, at least in a weak sense, can not be rejected for ASL flows just above a uniform bare soil surface when measurements are collected for "similar" mean meteorological conditions.

Other evidence of the validity of the ergodic hypothesis in the ASL is the success of Monin and Obukhov atmospheric surface layer similarity theory (MOST). MOST demonstrates that for unstable and near-neutral conditions, first and second moment velocity statistics from a wide range of ASL experiments collapse to few universal functions (to a first order). That is, "similar" mean meteorological conditions can be quantified in terms of mean surface heating and mean ground shear stress. Or, stated differently, differences in boundary conditions amongst those ASL experiments can be accounted for through normalization by surface heating

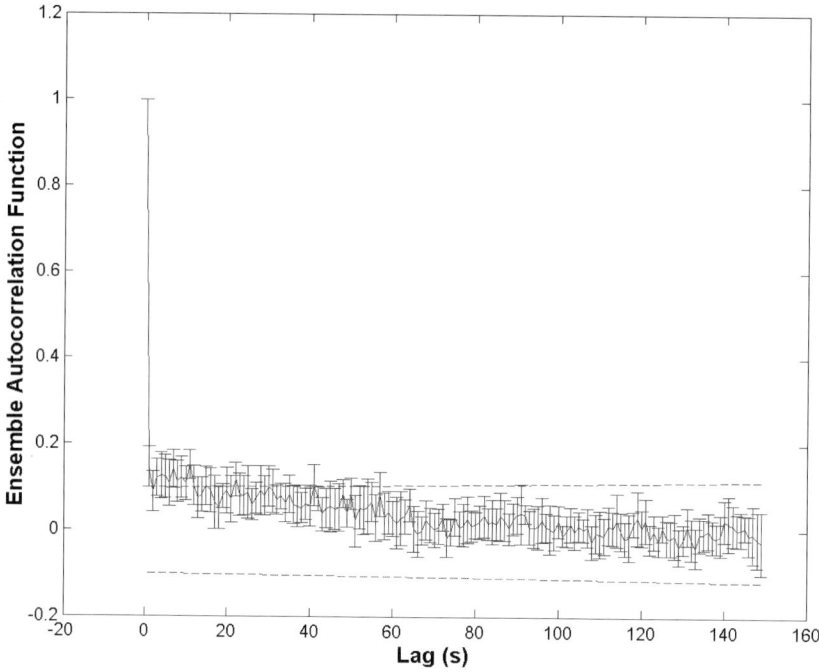

Figure 8.4. The ensemble autocorrelation function with time lag along with one standard deviation (vertical bar). The dotted lines are the 99% probability levels for which the time series is independent.

and surface shear stress. Certainly the success of similarity theory does not prove ergodicity, it still is a necessary condition for ergodicity in the ASL.

Recent LIDAR data analysis (Eichinger et al. 2001) goes one step further to demonstrate that the water vapor stability correction function (ϕ_v), computed from spatially distributed water vapor concentration profile measurements at the Campbell Tract facility in 1991, follow the recent theoretical ϕ_v derived from single tower measurements (Kader and Yaglom 1990). The difference between the tower-based estimates and the Eichinger evaluation of ϕ_v is that LIDAR based experiments generate an ensemble of ϕ_v collected under similar mean meteorological and hydrologic conditions, albeit at one site. The fact that temporal estimates of ϕ_v from numerous field experiments and spatial estimates of ϕ_v obtained by LIDAR converge suggests that the near-neutral and unstable ASL are, to a first approximation, stationary and ergodic in at least a weak sense close to the ground surface.

Higher in the ASL, entrainment of water vapor at the capping inversion can modulate the ASL turbulence by injecting low frequency modes leading to non-stationarity in time series of q (Mahrt 1998). That is, even if the mean meteorological states are similar close to the ground surface, differences in the entrainment fluxes among these runs can lead to non-stationarity. The validity of stationarity, homogeneity, and ergodicity in the stable ASL is much less certain but is beyond the scope of this study, which focuses on the CSL.

4 Homogeneity and Ergodicity in the Neutral and Unstable CSL

As earlier discussed, ergodicity is intimately linked to the definition of ensemble of experiments. The use of "similar" mean meteorological and hydrologic states (as described by the stability parameter) proved to be practical for evaluating stationarity, homogeneity, and ergodicity in the ASL. Hence, a logical question is whether the use of "similar" mean meteorological and hydrological states is equally complete for defining ensembles in the CSL.

Unfortunately, the answer to this question appears to be "no". The particular spatial variation in canopy roughness features can cause spatial variation of the statistics in addition to the effect from the mean surface heating and ground shear stress (both are not independent of these roughness features).

For example, at a given z, velocity statistics in the wake of an obstacle are distinct from velocity statistics more removed from the obstacle thereby demonstrating that planar homogeneity cannot exist at these elevations. Given that homogeneity is a necessary condition for ergodicity, it is clear that the CSL flow cannot be ergodic (in space). Stated differently, if the velocity measurements in the wake of the obstacle are sampled sufficiently long in time, they cannot capture all possible velocity values away from the obstacle. If, on the other hand, the experiment is repeated numerous times for the same setup, the velocity statistics in the wake of the obstacle and far from the obstacle are repeatable locally across experiments (Poggi et al. 2003). In short, if CSL and ASL "ensemble" definitions are taken to be identical, the CSL flow cannot be ergodic. On the other hand, if ensemble is defined for a given z and canopy geometry, then the flow might be treated as ergodic. This means that ergodicity may permit us to replace time averages with ensemble averages; however, spatially, ergodicity does not exist. If we seek to estimate ecosystem fluxes from a single tower, a new look at the definition of ensemble of experiments is required. Few practical "fixes" have been informally adopted for the lack of homogeneity (and hence ergodicity) in the CSL:

- On theoretical grounds, spatial averaging was proposed to augment temporal averaging (Wilson and Shaw 1977, Raupach and Shaw 1982). The basic premise here is that if we seek to measure (or model) average flow statistics that reflect horizontally average canopy morphology, then a spatial averaging of temporal statistics becomes logical. It is envisaged that the addition of such spatial averaging recovers "planar homogeneity" at scales much larger than the local spatial variation in flow statistics near and far from obstacles. That is, we assume that heterogeneity is confined to scales smaller than some critical scale. In the case of a laboratory setup in which the vegetation is composed of an array of equally spaced cylinders, such critical scale may be thought of as the "cell scale" between the cylinders (Poggi et al. 2003). In forested ecosystem, the precise estimate of such a critical scale is not clearly established.

- The moving equilibrium hypothesis or local similarity which accepts the view that the CSL is not planar homogeneous as the ASL, but perhaps is locally scalable (i.e. similarity relations) in terms of local parameters. This point has been explored in a CSL spatial variability experiment at the canopy-atmosphere interface of an even-aged pine forest (Katul et al. 1999). This experiment reported the spatial variability in scalar fluxes and velocity statistics from 7 towers. A key outcome was that while the 30 minute time averaged fluxes vary appreciably in space within this single even-aged canopy, the locally-scaled flux-variance relationships appeared more planar homogeneous. The moving equilibrium hypothesis is most accurate when the local horizontal gradients in the flow statistics are secondary or if the advective terms affect both fluxes and variances in the same direction so that upon normalization, cancelation of their effects occurs. For example, a "warm" patch in a forest would experience a higher sensible heat flux and a higher air temperature standard deviation. Hence, the flux-variance relationships across the forest are less variable across the forest-atmosphere interface when compared to their sensible heat flux counterpart.

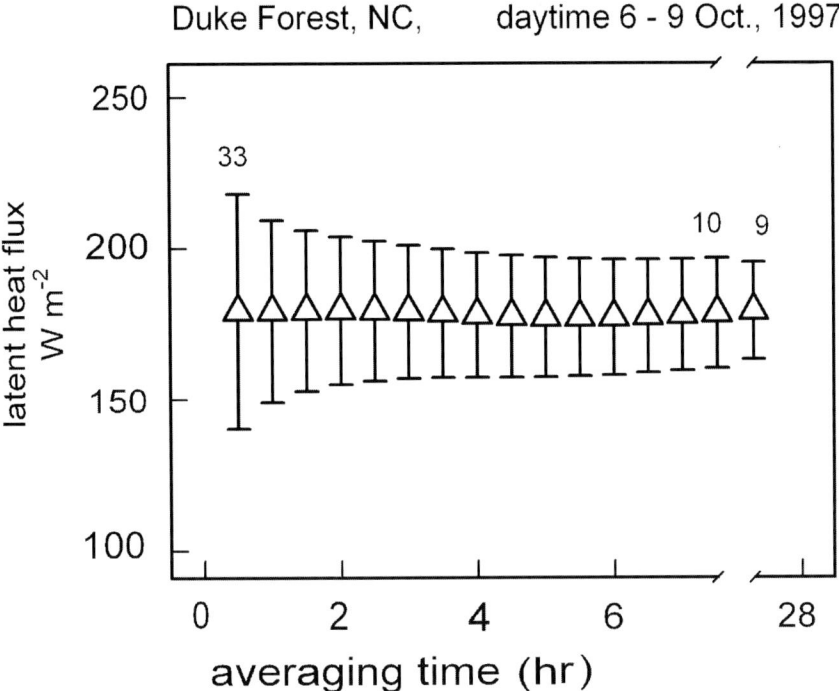

Figure 8.5. The decrease in spatial variability of daytime latent heat fluxes with increase in averaging interval. The minimum spatial variability obtained on a daily time averages are 9% while the reference 30 minute spatial variability is 33%.

- The natural variability in wind direction permits single towers to sample simultaneous spatial and temporal averages thereby minimizing the impact of planar non-homogeneity on the overall relationship between flow statistics and boundary conditions. The combination of spatial and temporal averages at one point may lead to the following hypothesis: By increasing the (temporal) averaging period, the spatial variability diminishes. Stated differently, as the averaging period increases, the spatial flow statistics resemble a planar-homogeneous CSL flow. This is expected because with increase in the sampling period (along with variable wind direction), the source-weight function contributing to a tower measurement becomes more representative of the entire stand. Hence, if the flow statistics are sampled at several points within the stand over sufficiently long periods of time, these flow statistics are likely to converge because of the overlap in source-weight functions. Indirect evidence supporting this hypothesis comes from the previously described CSL spatial variability field experiment (Katul et al. 1999). Again, in this experiment, spatial variation (standard deviation across the towers) in time averaged daytime latent heat flux decreased from 33% (at the reference 30-minute time scale) to 10% when averaged over 6 hours, as evidenced by Figure 8.5. Naturally, the result in Figure 8.5 is a necessary but not sufficient condition to validate the above hypothesis as other factors such as reduction in random errors by increased averaging interval significantly contribute.

Thus far, our focus was on the planar homogeneity of the unstable and near-neutral CSL. For the stable CSL, the lack of planar homogeneity is further compounded by the lack of potential stationarity described next.

5 Stationarity and Ergodicity in the Stable CSL

Stable CSL flows are complicated by numerous transient phenomena including intermittent non-turbulent processes such as meandering or wavy motion (Lee et al. 1997, Lee 1997, Lee and Barr 1998, Mahrt 1999, Hu et al. 2002). Several studies also reported synchronous occurrences of inverted ramps in the canopy and of canopy waves above, both characterized by comparable return periods (Paw et al. 1992, Lee et al. 1997, Lee 1997, Lee and Barr 1998, Mahrt 1999, Hu et al. 2002). Recently, Cava and co-workers (Cava et al. 2004) analyzed nighttime runs from the stable CSL of the Duke pine forest at $z/h = 1.12$ and found that:

- For near-neutral and slightly stable flows, canopy ramps dominate much of the momentum and scalar exchange process, including sensible heat, latent heat, and CO_2 fluxes (Figure 8.6).

- For very stable flows, linear gravity waves near the canopy top appear to dominate much of the variances, but the mean scalar fluxes across several "wave-periods" are small (Figure 8.7).

- The strong stability damping of turbulence appears to be disrupted by passage of clouds (see net radiation time series in Figure 8.8).

It is evident from Figure 8.8 that transients in the upper boundary conditions (e. g. passage of clouds) are one possible cause of the lack of stationarity (and hence ergodicity) for the stable CSL. Cava and co-workers also found that when the flow is not strongly stable and therefore dominated by ramp motion (analogous to conditions shown in Figure 8.6), the passage of clouds have an insignificant effect on the transport properties (see net radiation time series in Figure 8.6). The

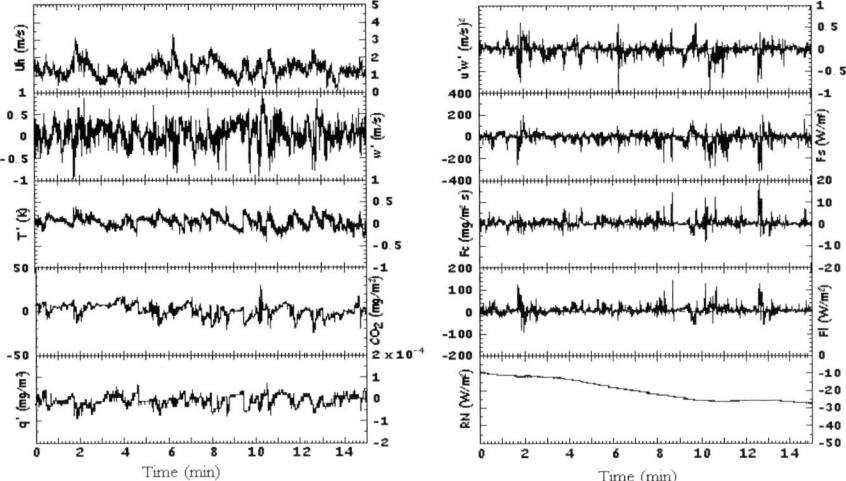

Figure 8.6. The temporal variation in longitudinal velocity (U_h), vertical velocity (w'), temperature (T'), CO_2, and water vapor concentration excursions (q') for mildly stable conditions (left panel). Fluxes of momentum ($u'w'$), sensible heat (F_s), CO_2 (F_c), and latent heat (F_l) are also shown (right panel). For reference the variation of net radiation (R_n) is shown as well.

consequences of the above three findings to nighttime CO_2 flux corrections are significant. Ecosystem respiration estimates collected for high u_* are typically dominated by ramp-like motion, analogous to that shown in Figure 8.6, and not overly sensitive to external perturbations such as passage of clouds. However, for low u_*, the intermittent switching between ramps, damped turbulence by strongly stable conditions, and no turbulence dominated by canopy waves can induce non-stationarity as suggested by the conceptual model of Figure 8.9. When the flow is dominated by canopy waves, small perturbations in radiative forcing or other instabilities leads to wave break-up thereby making the strongly stable CSL overly sensitive to the upper boundary condition (when compared to the unstable or near-neutral CSL). Furthermore, it is likely that gravity wave-turbulence interaction leads to the formation of non-linear waves (Finnigan 1999) that significantly transport CO_2 intermittently

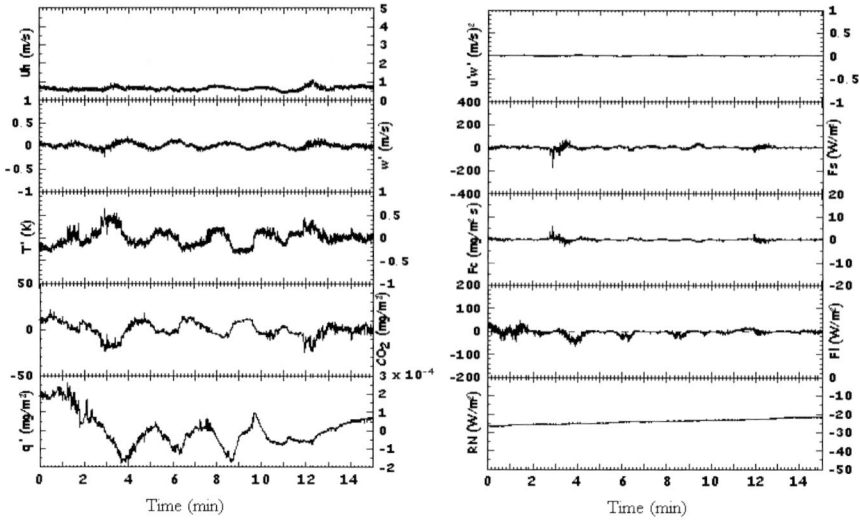

Figure 8.7. Same as Figure 8.6 but for very stable conditions.

within the CSL thereby compounding the non-stationarity. The above problem is even more amplified in open canopies, where the subcanopy becomes strongly stratified and de-coupled from ramps, waves, and other events above the canopy (Mahrt et al. 2000).

6 Conclusions

Stationarity, homogeneity, and ergodicity are routinely used to link turbulence field measurements collected in the ASL and CSL to land-surface processes. We showed that the near-neutral and unstable ASL are sometimes sufficiently stationary and the ensemble can be replaced by one realization. We also showed that extending the definition of the ASL ensemble to the CSL intuitively leads to non-homogeneous and non-ergodic flow properties. We showed that a few "fixes" such as spatial averaging of NS, the moving equilibrium hypothesis, or natural variation in wind direction when integrated over sufficiently long periods offer some promise to reducing the degree of non-homogeneity in the CSL. A more significant problem for CSL flows is stationarity (and ergodicity) during stable flows. We showed that when the flow is mildly stable, ramp-like

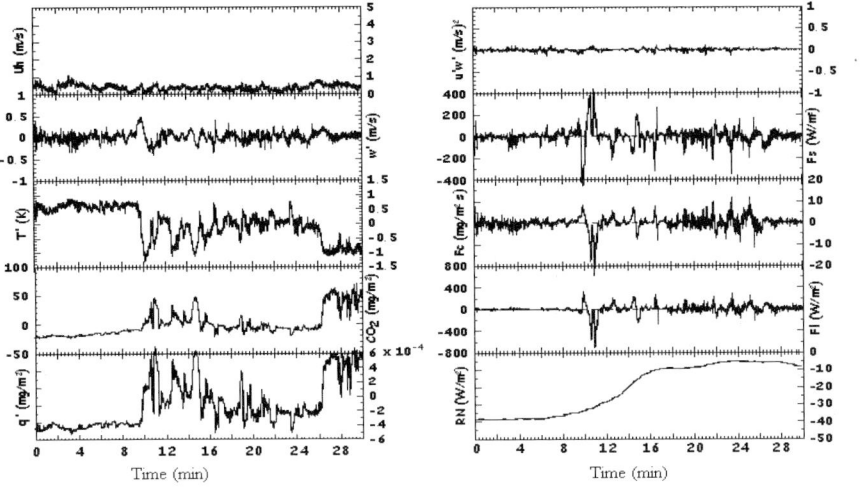

Figure 8.8. Same as Figure 8.6 but for strongly stable conditions before and after the passage of a cloud (note the increase in R_n).

motion dominate the exchange of momentum, mass, and energy between the land surface and the atmosphere. For such conditions, the flow appears to be stationary (at least for the data examined thus far). When the flow is extremely stable (no turbulence), the CSL is dominated by intermittent canopy waves. The intermittency of these waves, often subsets of "external" intermittency, is responsible for non-stationarity. The term "external intermittency" is used to distinguish such intermittency in large scale flow features from the classical "internal" intermittency often studied in the context of Kolmogorov scaling, local isotropy, and dissipation at fine scales. At the transition between the turbulent and non-turbulent states (i. e. very stable flow), the CSL flow appears overly sensitive to small radiative perturbations, such as passage of clouds. Again, these conditions result in non-stationary conditions within the CSL. Implications to monitoring nighttime CO_2 fluxes is that runs collected under near-neutral conditions must be selected to avoid such stationarity issues. Also, averaging over longer time periods across various wind directions is necessary for reducing the non-homogeneity inherent to all CSL flows.

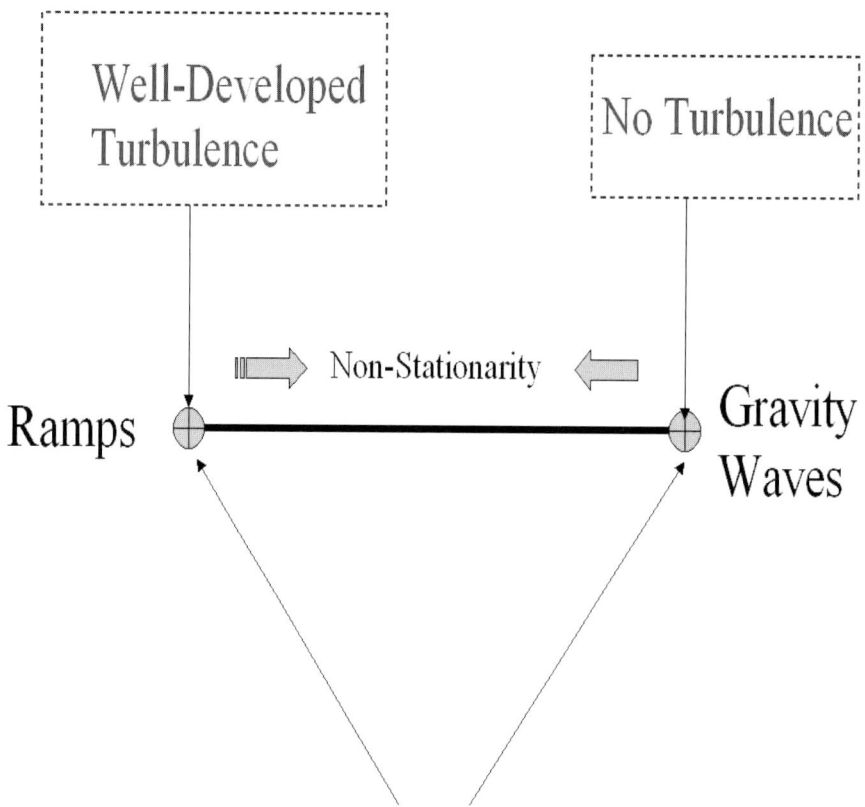

Figure 8.9. The two-end member states of the stable CSL and the organized motion dominating them. Both of these states are stationary; however, any shift from these two states leads to a non-stationary CSL flow.

7 Acknowledgment

Support was provided by the National Science Foundation (NSF-EAR and NSF-DMS), the Biological and Environmental Research (BER) Program, U. S. Department of Energy, through the Southeast Regional Center (SERC) of the National Institute for Global Environmental Change (NIGEC), and through the Terrestrial Carbon Processes Program (TCP) and the FACE project, and by the Italian MURST Project "Sviluppo di Tecnologie innovative e di processi biotecnologici in condizioni controllate nel settore delle colture vegetali: Diagnosi e Prognosi di situazioni di stress idrico per la vegetazione". The authors thank Ram Oren and Cheng-I Hsieh for their assistance in the analysis leading to Figure 8.5.

8 References

Anderson, O. D.: 1976, *Time Series Analysis and Forecasting: The Box-Jenkins Approach*, Butterworth, London.

Baldocchi, D., Finnigan, J. Wilson, K., Paw U, K. T., Falge, E.: 2000, 'On measuring net ecosystem carbon exchange over tall vegetation on complex terrain', *Bound.-Layer Meteorol.* **96**, 257-291.

Cava, D., Giostra, U., Siqueira, M. B. B., Katul, G. G: 2004, 'Organised motion and radiative perturbations in the nocturnal canopy sublayer above an even-aged pine forest', *Bound.-Layer Meteorol.* **112**, 129-157.

Eichinger, W. E., Cooper, D. I., Archuletta, F. L., Hof, D., Holtkamp, D. B., Karl, R. R., Quick, C. R., Tiee, J.: 1994, 'Development of a scanning, solar-blind, water raman Lidar', *Applied Optics*, **33**, 3923-3932.

Eichinger, W. E., Parlange, M. B., Katul, G. G.: 2001, 'Lidar measurements of the dimensionless humidity gradient in the unstable atmospheric surface layer', Pages 7-13 in V. Lakshmi, J. Albertson, and J. Schaake, editors. *Land Surface Hydrology, Meteorology, and Climate*, American Geophysical Union, Washington, D. C.

Finnigan, J.: 1999, 'A note on wave-turbulence interaction and the possibility of scaling the very stable boundary layer', *Bound.-Layer Meteorol.* **90**, 529-539.

Finnigan, J.: 2000, 'Turbulence in plant canopies', *Ann. Rev. Fluid Mechanics* **32**, 519-571.

Finnigan, J. J., Clement, R., Malhi, Y., Leuning, R., Cleugh, H. A.: 2003, 'A re-evaluation of long-term flux measurement techniques. Part I: Averaging and coordinate rotation', *Bound.-Layer Meteorol.* **107**, 1-48.

Finnigan, J. J., Raupach, M. R., Bradley, E. F., Aldis, G. K.: 1990, 'A wind-tunnel study of turbulent flow over a 2-dimensional ridge', *Bound.-Layer Meteorol.* **50**, 277-317.

Frisch, U.: 1995, *Turbulence*, Cambridge University Press, Cambridge.

Giostra, U., Cava, D., Schipa, S.: 2002, 'Structure functions in a wall-turbulent shear flow', *Bound.-Layer Meteorol.* **103**, 337-359.

Hu, X. Z., Lee, X., Stevens, D. E., Smith, R. B.: 2002, 'A numerical study of nocturnal wavelike motion in forests', *Bound.-Layer Meteorol.* **102**, 199-223.

Kader, B. A., Yaglom, A. M.: 1990, 'Mean fields and fluctuating moments in unstably stratified turbulent boundary layers', *J. Fluid Mechanics* **212**, 637-662.

Kaimal, J. C., Finnigan, J. J.: 1994, *Atmospheric Boundary Layer Flows: Their Structure and Measurement*, Oxford University Press, New York.

Katul, G., Hsieh, C. I., Bowling, D., Clark, K., Shurpali, N., Turnipseed, A., Albertson, J., Tu, K., Hollinger, D., Evans, B., Offerle, B., Anderson, D., Ellsworth, D., Vogel, C., Oren, R.: 1999, 'Spatial variability of turbulent fluxes in the roughness sublayer of an even-aged pine forest', *Bound.-Layer Meteorol.* **93**, 1-28.

Katul, G., Vidakovic, B., Albertson, J.: 2001, 'Estimating global and local scaling exponents in turbulent flows using discrete wavelet transformations', *Physics of Fluids*, **13**, 241-250.

Lee, X., Neumann, H. H., Den Hartog, G., Fuentes, J. D., Black, T. A., Mickle, R. E., Yang, P. C., Blanken, P. D.: 1997, 'Observation of gravity waves in a boreal forest', *Bound.-Layer Meteorol.* **84**, 383-398.

Lee, X. H.: 1997, 'Gravity waves in a forest: A linear analysis', *J. Atmos. Sci.* **54**, 2574-2585.

Lee, X. H., Barr, A. G.: 1998, 'Climatology of gravity waves in a forest', *Quart. J. R. Meteorol. Soc.* **124**, 1403-1419.

Lesieur, M.: 1990, *Turbulence in Fluids*, 2 edition. Kluwer Academic Publishers, Boston.

Mahrt, L.: 1998, 'Flux sampling errors for aircraft and towers', *J. Atmos. Oceanic Techn.* **15**, 416-429.

Mahrt, L.: 1999, 'Stratified atmospheric boundary layers', *Bound.-Layer Meteorol.* **90**, 375-396.

Mahrt, L., Lee, X. H., Black, A., Neumann, H., Staebler, R. M.: 2000, "Nocturnal mixing in a forest subcanopy', *Agric. For. Meteorol.* **101**, 67-78.

Monin, A. S., Yaglom, A. M.: 1971, *Statistical Fluid Mechanics*, The MIT Press, Cambridge.

Paw U, K. T., Brunet, Y., Collineau, S., Shaw, R. H., Maitani, T., Qiu, J., Hipps, L.: 1992, 'On Coherent Structures in Turbulence above and within agricultural plant canopies', *Agricul. For. Meteorol.* **61**, 55-68.

Poggi, D., Porporato, A., Ridolfi, L., Albertson, J., Katul, G. G.: 2003, 'The effect of vegetation density on canopy sublayer turbulence', *Bound.-Layer Meteorol.* in review.

Raupach, M. R., Finnigan, J. J.: 1997, 'The influence of topography on meteorological variables and surface-atmosphere interactions', *J. Hydrol.* **190**, 182-213.

Raupach, M. R., Shaw, R. H.: 1982, 'Averaging procedures for flow within vegetation canopies', *Bound.-Layer Meteorol.* **22**, 79-90.

Stanisic, M.: 1985, *Mathematical Theory of Turbulence*, Springer-Verlag, New York.

Tennekes, H., Lumley, J. L.: 1972, *A First Course in Turbulence*, Massachusetts Institute of Technology, Boston.

Wilson, N. R., Shaw, R. H.: 1977, 'A higher order closure model for canopy flows', *J. Applied Meteorol.* **16**, 1197-1205.

Chapter 9

POST-FIELD DATA QUALITY CONTROL

Thomas Foken, Mathias Göckede, Matthias Mauder, Larry Mahrt,
Brian Amiro, William Munger
thomas.foken@uni-bayreuth.de

Abstract This Chapter summarizes the steps of quality assurance and quality control of flux measurements with the eddy covariance method. An important part is the different steps of the control for electronic, meteorological and statistical problems. The fulfillment of the theoretical assumptions of the measuring method and the non-steady state test and the integral turbulence test are extensively discussed as well as an overall flagging for data quality and a site specific quality analysis using footprint models. Finally, problems are discussed which are not included yet in the control program, mainly connected with the complicated turbulence structure at a forest site.

1 Introduction

A consistent procedure for quality control of meteorological data is essential for measurement networks and long-term measurement sites. This issue has been extensively addressed for standard meteorological networks. Reliable, automated procedures based on inspection of time series which can reduce quality control efforts and provide a consistent product across measurement networks, have been the focus of several studies. Smith et al. (1996) have constructed automated quality control procedures for slow response surface data that flag questionable data points for visual inspection. Hall et al. (1991) examined the quality assurance of observations from ships and buoys using output from a numerical weather prediction model as a constraint. Lorenc and Hammon (1988) constructed an automated procedure to flag errors from ship reports, buoys and synoptic reports. They concluded that their procedure does not give completely reliable results, and that subjective analysis did better than the automated program during unusual conditions, such as developing depressions. Essenwanger (1969) presented an automated procedure for detecting erroneous or suspicious observational records based on obvious data errors, comparison of adjacent (in time or space) data, and by comparing to prescribed limits of a standard Weibull distribution. Essenwanger (1969) concluded that his automated technique could not unequivocally pinpoint differences between a rare event and an instrument problem. DeGaetano (1997) presents a scheme to quality control wind measurements. Methods to control radiation measurements were discussed by Gilgen et al. (1994), which can be implemented into continuously running systems.

In contrast to standard meteorological measurements there are only a few papers available that discuss quality control of eddy covariance measurements (Foken and Wichura 1996, Vickers and Mahrt 1997). Quality control of eddy covariances should include not only tests for instrument errors and problems with the sensors, but also evaluate how closely conditions fulfill the theoretical assumptions underlying the method. Because the latter depends on meteorological conditions, eddy covariance quality control tools must be a combination of a typical test for high resolution time series and examination of the turbulent conditions. A second problem is connected with the representativity of the measurements depending on the footprint of the measurement. The control of the percentage of the area of interest in the actual footprint is a further issue. It is the aim of the present Chapter to describe a set of possible tests and protocol for data flagging and give practical guidance for use in continuously running eddy covariance systems like the FLUXNET program.

2 Quality Assurance and Quality Control

Quality assurance is one of the most important issues for creation and management of a measuring program. Issues of quality assurance are widely known for routine meteorological measuring programs (Shearman 1992). The present network of carbon dioxide flux sites evolved from an assemblage of individual sites with varying objectives (biological or micrometeorological) and protocols, rather than being designed from the outset as a network. Therefore, the quality assurance of such measuring programs was written after the measurements had started (e. g. Aubinet et al. 2000, Moncrieff et al. 1997). And even now some of the topics are under discussion. A quality assurance (QA) scheme needs the following components:

- Specification of user requirements: The users of the flux data, which may be modelers or policy-makers, who need the information for example in the Kyoto process, need basic information of the measuring program such as accuracy, resolution in time and space (number of sites and surface types). An important task is the development of reliable and feasible measuring programs.

- Specification of the measuring system: A suitable measuring system must be developed according to the requirements and the personal, financial and scientific constraints. This was partly done (Moncrieff et al. 1997), but presently different types of systems are used because of changes and improvements in the measuring technique. This makes the comparability of the results of different sets of instruments difficult and comparison experiments are urgently required.

- Identification of suitable measuring locations: This is a most difficult problem, because several measuring stations were created where research facilities were already in place, rather than being selected according to micrometeorological criteria. Therefore, site characterization tools are needed to ensure data quality (see Section 3.3). Ideally, site selection would be made based on quality testing of data collected from a temporary tower prior to construction of an expensive tower station.

- Definition of necessary calibrations: Calibrations allow comparison of data among sites. The accuracy of any measurement is ultimately limited by the accuracy and frequency of calibration standards that are used. Most of the necessary calibrations and control issues are well described (e. g. Aubinet et al. 2000, Goulden et al. 1996, Moncrieff et al. 1997).

- Definition of quality control (QC): The most important part of quality assurance is quality control. Several tests are discussed in this Chapter. Quality control must be done in realtime or shortly after the measurements to minimize data loss by reducing the time to detect and fix instrument problems.

- Quality evaluation: This topic is similar to QC. The main difference is a description of the data quality to be able to compare data for different periods and sites. This is also a main goal of the present Chapter.

- Corrective actions: Corrective actions refers to corrections caused by calibrations, by the choice of the coordinate system, and the sensor size and separation, etc. Most of the corrections are discussed in the other Chapters of the book and the literature (Aubinet et al. 2000, Moncrieff et al. 1997, etc.).

- Feedback from the user of the data: The database is often the end product of a measuring program. However, the user needs some control of the data and the opportunity to provide feedback to the experimentalist to improve the data quality and to make necessary changes in the program.

3 Quality Control of Eddy Covariance Measurements

A uniform scheme does not exist for quality control of eddy covariance measurements. Only several aspects are discussed in the literature. For the producer of flux data there are a number of specific techniques but no instructions for practical handling of the data. In the following, an overview of different quality control steps is given:

- The first steps of data analysis are basic tests of the raw data (Vickers and Mahrt 1997) such as electrical tests of the amplitude, the resolution of the signal, the control of the electronic and meteorological range of the data and spikes (Højstrup 1993), which are discussed further in Section 3.1.

- Statistical tests must be applied to sampling errors of the time series (Finkelstein and Sims 2001, Haugen 1978, Vickers and Mahrt 1997) and are discussed in Section 3.2. Also abrupt step changes in the time series, or reasons for non-stationarity must be identified (Mahrt 1991, Vickers and Mahrt 1997).

- A main issue for quality control are tests on fulfillment of the requirements for eddy covariance measurements. Steady state conditions and a developed turbulent regime are influenced not from the sensor configuration but from the meteorological conditions (Foken and Wichura 1996). The fulfillment of these conditions is discussed in Section 3.3.

- A system of general quality flagging of the data is discussed in Section 3.4 and a site specific evaluation of the data quality using footprint models is in Section 3.5.

3.1 Basic tests of the raw data

Vickers and Mahrt (1997) developed a framework of test criteria for quality control of fast response turbulence time series data with a focus on turbulent flux calculations. The tests are not framed in terms of similarity theory, nor do they assume that the fields necessarily follow any particular statistical distribution. Many types of instrument malfunctions can be readily identified with simple automated criteria. However, even after tuning the threshold values, the automated tests still occasionally identify behaviors that appears to be physical after visual inspection. Physically plausible behavior and instrument problems can overlap in parameter space. This underscores the importance of the visual inspection step in quality control to either confirm or deny flags raised by the automated set of tests. Data flagged but later deemed physical after graphical inspection are often found to be the most unusual and interesting situations, including intermittent turbulence, downward turbulence bursting, microfronts, gravity waves and other stable boundary layer phenomena. Some automated tests for quality control of turbulence time series are briefly summarized below.

Spikes are typically characterized as short duration, large amplitude fluctuations that can result from random noise in the electronics (Brock 1986). Quality control should include the identification and removal of spikes. For example, correlated spikes in the temperature and vertical velocity from a sonic anemometer can contaminate the calculated heat flux. Spikes that do not influence the fluxes still affect the variances. When the number of spikes becomes large, the entire data period should be considered suspect and discarded. The effect of water collecting on the transducers of some sonic anemometers often appears as spikes. Less than optimum electrical power supplies, which are sometimes necessary at remote measurement sites, can lead to frequent spiking. Unrealistic data values occur for a number of reasons. These data should be detected by comparing the minimum and maximum values to prescribed limits. For example, a vertical velocity in excess of 5 m s^{-1} close to the ground is probably not physical. However, visual inspection is sometimes required due to special circumstances, such as high turbulence levels associated with exceptionally strong surface heating. Højstrup (1993) tested a data screening procedure for application to Gaussian distributed turbulence data. Spikes are absolute quantities of measuring values which are larger than approximately four times of the standard deviation of the time series. This test should be repeated 2 or 3 times with each time series.

Some success identifying instrument problems has been achieved by comparing higher moment statistics to threshold values. Abnormally large skewness often indicates a problem, although care must be taken because, for example, the temperature near the ground during strong surface heating typically has large positive skewness. Unusually small or large kurtosis often indicates an instrument problem. Large kurtosis in the temperature field from a sonic anemometer is sometimes related to spiking associated with water on the transducers. Most despiking algorithms fail to remove this persistent type of spiking, in contrast to short duration high amplitude spikes associated with noise in the electronics. Histograms of values of a single turbulence channel are also useful. A non-typical distribution of the measuring data can indicate averaging errors connected with the digitization. Such errors were found for the Solent sonic anemometers R2 and R3 (Chr. Thomas, University of Bayreuth, 2002, personal communication, problem solved partly by Gill in 2003). In this case for example the R2 measured no vertical wind of -0.01 $\mathrm{m\,s^{-1}}$ but the number of measuring points for 0.00 $\mathrm{m\,s^{-1}}$ was twice as high as the other data. This indicates a small shift to positive vertical wind velocities.

Unusually large discontinuities in the mean can be detected using the Haar transform. The transform is simply the difference between the mean calculated between two adjacent windows. Large values of the transform identify changes in the mean that are coherent on the time scale of the window width. The goal here is to detect semi-permanent changes as opposed to smaller scale fluctuations. A sudden change of offset is one example of an instrument related jump in mean variables. The window size and the threshold values that identify suspect periods may need adjustment for particular datasets. For example, for aircraft data in the convective boundary layer, the mean vertical wind may change significantly as the aircraft enters and exits large scale coherent thermals. However, for tower measurements close to the ground, coherent changes in the mean vertical wind are typically much smaller. Care must be taken with aircraft data over heterogeneous surfaces, where coherent changes in the mean fields are common due to the formation of local internal boundary layers. For example, a sharp change in mean temperature will be found where the aircraft intersects the top of a warm internal boundary layer. In less clear cases, data from other levels and other instruments should be consulted for verification.

Instrument problems can also be detected by comparing the variance to prescribed thresholds. A sequence of variances should be calculated for a sequence of sliding, overlapping windows to detect isolated problems. For example, a brief period with near zero temperature fluctuations could be due to a temporarily non-responding instrument. Visual inspection is sometimes necessary in stable conditions where the true physical variances can become very small, usually due to a combination of strong temperature stratification and weak mean wind shear. Unusually large variance often indicates an instrument malfunction.

In recent years many closed path carbon dioxide analyzers (LiCor 6262) were replaced by open path sensors (LiCor 7500). These sensors are more sensitive to rain and frost. The development of a site-specific test using precipitation, radiation wind and temperature data can help to indicate these situations. This can be done with statistical methods like multiple regressions. Such tests can be important, because interference is not always clearly indicated in the time series.

3.2 Statistical tests

The calculation of means, variances and covariances in geophysical turbulence is inherently ambiguous, partly due to nonturbulent motions on scales which are not large compared to the largest turbulent eddies. As a result of these motions, geophysical time series are normally nonstationary to some degree (Foken and Wichura 1996, Vickers and Mahrt 1997). The physical interpretation of the flux computed from nonstationary time series is ambiguous in that it simultaneously represents different conditions and the computed perturbations for calculation of the flux are contaminated by nonstationarity, which can only be partially removed by detrending or filtering. Nonturbulent motions contaminate the flux calculation in that the flux due to nonturbulent motions may be primarily random error, as found in Sun et al. (1996). Attempts to remove nonstationarity by trend removal or filtering violates Reynolds averaging, although often the errors are small. Attempts to reduce the nonstationarity by reducing the record length increases the random flux error. Techniques for approximately separating random variations and nonstationarity are presented in Mahrt (1998) and Trevino and Andreas (2000). Tests on non-steady state conditions are given in Section 3.3.1.

Systematic errors (flux bias) result from failure to capture all of the turbulent transporting scales (Foken and Wichura 1996, Lenschow et al. 1994, Oncley et al. 1996, Vickers and Mahrt 1997). Such systematic errors occur at either the large scale end where the largest transporting eddies may be excluded from the flux calculation, or at the small scale end where transport by small eddies can be eliminated by instrument response time, pathlength averaging, instrument separation and post-process filtering. With weak winds and substantial surface heating, many flux calculation procedures may exclude larger-scale turbulent flux due to slowly moving boundary-layer scale eddies (Sakai et al. 2001). Increasing the averaging time also captures nonturbulent, mesoscale motions (nonstationarity). With very stable conditions, turbulence quantities may be confined to very short time scales, sometime less than one minute (Vickers and Mahrt 2003). Use of traditional averaging periods of five minutes or more leads to perturbation quantities, which are strongly contaminated by gravity waves, meandering motions and other mesoscale motions (see Mahrt et al. 2001a and references therein). Some of these problems can be identified with the tests given in Section 3.3.2.

The random flux error is the uncertainty due to inadequate record length and the random nature of turbulence (Finkelstein and Sims 2001, Lenschow et al. 1994, Lumley and Panofsky 1964, Mann and Lenschow 1994, Vickers and Mahrt 1997). Once perturbation quantities are computed and products are taken to compute variances, fluxes and other turbulence moments, the turbulence quantities can be averaged over a longer time period to reduce random sampling errors. The latter is sometimes referred to as the "flux-averaging time scale" to distinguish it from the shorter averaging time scale used to define the perturbations. The time scale for averaging the flux normally should be longer than that used to compute the perturbations themselves. Reynolds averaging can still be satisfied as long as the averaging is unweighted (no filtering or detrending) (Mahrt et al. 2001b). For example, one might choose an averaging time of 2 minutes for very stable conditions but wish to average the 2-minute fluxes over 30 minutes or one hour to reduce random flux errors.

With very stable conditions where the turbulence is intermittent, reduction of the random error to acceptable levels may require a prohibitively long averaging time (e. g. Haugen 1973). The flux for a one-hour period can be dominated by one or two events and therefore a much longer averaging time is required. Howell and Sun (1999) choose the record length by attempting to objectively maximize the flux and minimize the random flux error.

The above results also apply to analysis of turbulence quantities from moving platforms such as aircraft, except that one must determine the averaging length from which to compute perturbations (often chosen to be 1 km) and choose the flux averaging length, sometimes chosen as the flight path length. In convective conditions with deep boundary layers, such an averaging length may exclude significant flux (Betts et al. 1990, Desjardins et al. 1992). The nonstationarity problem above becomes the heterogeneity problem for moving platforms (e. g. Desjardins et al. 1997). Reduction of random flux errors is facilitated by long flight paths for homogeneous surfaces or many repeated passes over heterogenous surfaces (Mahrt et al. 2002).

The autocovariance analysis is widely used to determine the time lag for closed-path gas analyzers (Leuning and Judd, 1996), because the concentration signal is measured some seconds later than the wind signal. Even data from open-path gas analyzer may have a small time offset between the measuring time and the position of the value in the data file because of electronic delays in recording and storing the data and finite signal processing times. If this is not known and not corrected in the logger program, it must be included in calculation of the fluxes. It is important to check the whole measuring system with an autocovariance analysis to identify time shifts between the signals.

3.3 Tests on fulfillment of theoretical requirements

The widely used direct measuring method for turbulent fluxes is the eddy covariance method, which involves a simplification of turbulent conservation equations for momentum and scalar fluxes, e. g., the flux of a scalar, c

$$F_c = \overline{w'c'} = \frac{1}{N-1} \sum_{k=0}^{N-1} [(w_k - \overline{w})(c_k - \overline{c})] \tag{9.1}$$

where w is the vertical wind component. This equation implies steady-state conditions. The choice of averaging length depends on the cospectra of the turbulence and steady state conditions. With an ogive test (Oncley et al., 1990)

$$Og_{w,c}(f_o) = \int_{-\infty}^{f_o} Co_{w,c}(f) df \tag{9.2}$$

where Co is the cospectra of the vertical wind velocity and the concentration. The convergence of Og at low frequencies indicates that all relevant eddies are collected. On the other hand an excessive measuring length may include nonsteady-state conditions (see Chapters 2 and 5). Therefore, these conditions should be tested for each time series, because they can influence the data quality significantly (see Section 3.3.1). However, in most cases, convergence occurs within a 30-minute period.

The integral turbulence characteristics in the surface layer may depend on the latitude (Johansson et al. 2001); this may be relevant for tests on eddy covariance measurements. The influence of density fluctuations can be corrected (see Chapters 6 and 7). Conditions of horizontal homogeneity must also be fulfilled in order to avoid significant advection, which can be influenced by the choice of the coordinate rotation (see Chapters 3 and 10).

Of greater importance is whether developed turbulent conditions exist, with very weak turbulence the measuring method and methods based on surface layer similarities may not be valid. Examination of normalized standard deviations (integral turbulence characteristics, see Section 3.3.2) provides an effective test for adequately developed turbulence. These tests are also sensitive to other influences on the data quality like limitations of the surface layer height, gravity waves, internal boundary layers, flow distortion, high frequency flux loss (see Chapter 4). For example, internal boundary layers and flow distortion problems of the sensors and towers can indicate higher standard deviations of turbulence parameters. For situations with gravity waves the correlation coefficient between the vertical wind velocity and scalars can be high, resulting in unusually large fluxes. Such situations, often during the night and under stable conditions, must be indicated and the wave and the turbulent signal must be separated (Handorf and Foken 1997).

Foken and Wichura (1996) applied criteria to fast-response turbulence data to test for non-stationarity and substantial deviations from flux-variance similarity theory, whether due to instrumental or physical causes. These are described below.

3.3.1 Steady state tests

Steady state conditions means that all statistical parameters do not very in time (e. g., Panofsky and Dutton, 1984). Typical non-stationarity is driven by the change of meteorological variables with the time of the day, changes of weather patterns, significant mesoscale variability, or changes of the measuring point relative to the measuring events such as the phase of a gravity wave. The latter may occur because of changing footprint areas, changing internal boundary layers (especially internal thermal boundary layers in the afternoon), or by gravity waves. Presently there are two main tests used to identify non-steady state conditions. The first is based on the trend of a meteorological parameter over the averaging interval of the time series (Vickers and Mahrt, 1997) and the second method indicates non-steady state conditions within the averaging interval (Foken and Wichura, 1996).

Vickers and Mahrt (1997) regressed the meteorological element x over the averaging interval of a time series and determined the difference of x between the beginning and the end of the time series according to this regression, δx. With this calculation they determined the parameter of relative non-stationarity, mainly for wind components

$$\text{RN}_x = \frac{\delta x}{\overline{x}} \qquad (9.3)$$

Measurements made over the ocean exceeded the threshold ($\text{RN}_x > 0.50$) 15 % of the time and measurements over forest exceeded the threshold

55 % of the time. A more rigorous measure of stationarity can found in Mahrt (1998).

The steady state test used by Foken and Wichura (1996) is based on developments of Russian scientists (Gurjanov et al., 1984). It compares the statistical parameters determined for the averaging period and for short intervals within this period. For instance, the time series for the determination of the covariance of the measured signals w (vertical wind) and x (horizontal wind component or scalar) of about 30 minutes duration will be divided into $M = 6$ intervals of about 5 minutes. N is the number of measuring points of the short interval ($N = 6{,}000$ for 20 Hz scanning frequency and a 5 minute interval):

$$(\overline{x'w'})_i = \frac{1}{N-1}\left[\sum_j x_j w_j - \frac{1}{N}\sum_j x_j \sum_j w_j\right]$$

$$\overline{x'w'} = \frac{1}{M}\sum_i (\overline{x'w'})_i \qquad (9.4)$$

This value will be compared with the covariance determined for the whole interval:

$$(\overline{x'w'})_o = \frac{1}{M(N-1)}\left[\sum_i(\sum_j x_j w_j)_i - \frac{1}{MN}\sum_i(\sum_j x_j \sum_j w_j)_i\right] \qquad (9.5)$$

The authors proposed that the time series is steady state if the difference between both covariances

$$\mathrm{RN_{COV}} = \left|\frac{(\overline{x'w'}) - (\overline{x'w'})_o}{(\overline{x'w'})_o}\right| \qquad (9.6)$$

is less than 30%. This value is found by long experience and is in a good agreement with other test parameters also of other authors (Foken and Wichura, 1996).

3.3.2 Test on developed turbulent conditions

Flux-variance similarity is a good measure to test the development of turbulent conditions. This similarity means that the ratio of the standard deviation of a turbulent parameter and its turbulent flux is nearly constant or a function of stability. These so-called integral turbulence characteristics are basic similarity characteristics of the atmospheric turbulence (Obukhov 1960, Wyngaard et al. 1971) and are routinely discussed in boundary layer and micrometeorology textbooks (Arya 2001, Foken 2003, Kaimal and Finnigan 1994, Stull 1988). Foken and Wichura

Table 9.1. Coefficients of the integral turbulence characteristics (Foken et al. 1997, Foken et al. 1991, Thomas and Foken 2002).

Parameter	z/L	c_1	c_2
σ_w/u_*	$0 > z/L > -0.032$	1.3	0
	$-0.032 > z/L$	2.0	1/8
σ_u/u_*	$0 > z/L > -0.032$	2.7	0
	$-0.032 > z/L$	4.15	1/8
σ_T/T_*	$0.02 < z/L < 1$	1.4	$-1/4$
	$0.02 > z/L > -0.062$	0.5	$-1/2$
	$-0.062 > z/L > -1$	1.0	$-1/4$
	$-1 > z/L$	1.0	$-1/3$

Table 9.2. Coefficients of the integral turbulence characteristics for wind components under neutral conditions (Thomas and Foken 2002).

Parameter	$-0.2 < z/L < 0.4$
σ_w/u_*	$0.21\ln(\frac{z_+ \times f}{u_*}) + 3.1,\ z_+ = 1\text{ m}$
σ_u/u_*	$0.44\ln(\frac{z_+ \times f}{u_*}) + 6.3,\ z_+ = 1\text{ m}$

(1996) used functions determined by Foken et al. (1991). These functions depend on stability and have the general form for standard deviations of wind components

$$\frac{\sigma_{u,v,w}}{u_*} = c_1 \left(\frac{z}{L}\right)^{c_2} \quad (9.7)$$

where u is the horizontal or longitudinal wind component, v the lateral wind component, u_* the friction velocity and L the Obukhov length. For scalar fluxes the standard deviations are normalized by their dynamical parameters (e. g., the dynamic temperature T_*)

$$\frac{\sigma_x}{X_*} = c_1 \left(\frac{z}{L}\right)^{c_2} \quad (9.8)$$

The constant values in Equations 9.7 and 9.8 are given in Table 9.1. For the neutral range the external forcing assumed by Johansson et al. (2001) and analyzed for the integral turbulence characteristics by Thomas and Foken (2002) was considered in Table 9.2 with the latitude (Coriolis parameter f). The parameters given for the temperature can be assumed for most of the scalar fluxes. It must be mentioned that under nearly neutral conditions the integral turbulence characteristics of the scalars have extremely high values (Table 9.1) and the test fails.

Table 9.3. Typical values for the correlation coefficient of the momentum and sensible heat flux.

Author	r_{uw}	r_{wT}
Hicks (1981)	-0.32	0.35 ($z/L \to -0.0$)
		0.6 ($z/L \to -2.0$)
Kaimal et al. (1990)	-0.3	0.5 ($z/L < 0.0$)
Kaimal and Finnigan (1994)	-0.35	0.5 ($-2 < z/L < 0$)
		-0.4 ($0 < z/L < 1$)
Arya (2001)	-0.15	0.6 ($z/L < 0.0$)

The test can be done for the integral turbulence characteristics of both parameters used to determine the covariance. The measured and the modeled parameters according to Equations 9.7 or 9.8 will be compared according to

$$\text{ITC}_\sigma = |\frac{(\sigma_x/X_*)_{\text{model}} - (\sigma_x/X_*)_{\text{measurement}}}{(\sigma_x/X_*)_{\text{model}}}| \quad (9.9)$$

If the test parameter ITC_σ is < 30 %, a well developed turbulence can be assumed.

A similar parameter is the correlation coefficient between the time series of two turbulent parameters. If this correlation coefficient is within the usual range (Table 9.3) a well-developed turbulence can be assumed (Kaimal and Finnigan, 1994).

3.4 Overall quality flag system

To be useful, the results of data quality checking must be made available in the final data archive. Measurements are normally flagged according to their status such as uncontrolled, controlled, corrected, etc. The quality tests given above open the possibility to flag also the quality of a single measurement. Foken and Wichura (1996) proposed to classify the tests according to Equations 9.6 and 9.9 into different steps and to combine different tests. An important parameter, which must be included in the classification scheme, is the orientation of the sonic anemometer, if the anemometer is not an omnidirectional probe and the measuring site does not have an unlimited fetch in all directions. For these three tests the definition of the flags is given in Table 9.4. Further tests, like an acceptable range of the mean vertical wind velocity, can be included into this scheme.

The most important part of a flag system is the combination of all flags into a general flag for easy use. This is done in Table 9.5 for the

Table 9.4. Classification of the data quality by the steady state test according to Equation 9.6 and the integral turbulence characteristics according to Equation 9.9 and the horizontal orientation of a sonic anemometer of the type CSAT3 (Foken 2003).

a		b		c	
class	range	class	range	class	range
1	0-15%	1	0-15%	1	±0-30°
2	16-30%	2	16-30%	2	±31-60°
3	31-50%	3	31-50%	3	±61-100°
4	51-75%	4	51-75%	4	±101-150°
5	76-100%	5	76-100%	5	±101-150°
6	101-250%	6	101-250%	6	±151-170°
7	251-500%	7	251-500%	7	±151-170°
8	501-1000%	8	501-1000%	8	±151-170°
9	>1000%	9	>1000%	9	> ±171°

a: State-state test according to Equation 9.6.
b: Integral turbulence characteristics according to Equation 9.9.
c: Horizontal orientation of the sonic anemometer.

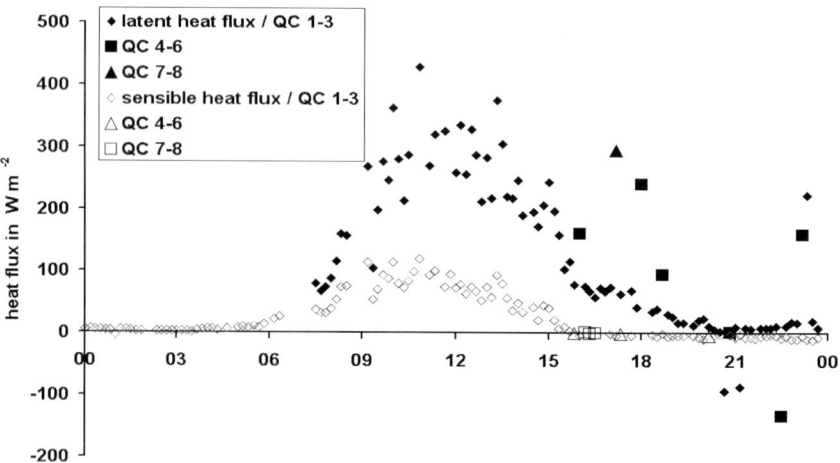

Figure 9.1. Daily cycle of the sensible and latent heat flux with quality classes measured by the University of Bayreuth during the LITFASS-1998 experiment (Beyrich et al. 2002) on June 02, 1998 in Lindenberg/Germany over grassland.

Table 9.5. Proposal for the combination of the single quality flags into a flag of the general data quality (Foken 2003).

a	b	c	d
1	1	1-2	1-5
2	2	1-2	1-5
3	1-2	3-4	1-5
4	3-4	1-2	1-5
5	1-4	3-5	1-5
6	5	≤ 5	1-5
7	≤ 6	≤ 6	≤ 8
8	≤ 8	≤ 8	≤ 8
9	*	*	*

a: Flag of the general data quality
b: Steady state test according to Equation 9.6
c: Integral turbulence characteristics according to Equation 9.9
d: Horizontal orientation of the sonic anemometer
*: One or more of flags b, c and d equals 9

flags given in Table 9.4. The user of such a scheme must know the appropriate use of the flagged data. The presented scheme was classified (by micrometeorological experiences) so classes 1 to 3 can be used for fundamental research, such as the development of parameterizations. The classes 4-6 are available for general use like for continuously running systems of the FLUXNET program. Classes 7 and 8 are only for orientation. Sometimes it is better to use such data instead of a gap filling procedure, but then these data should not differ significantly from the data before and after these data in the time series. Data of class 9 should be excluded under all circumstances. Such a scheme gives the user a good opportunity to use eddy covariance data. Finally the data can be presented together with the quality flag like in Figure 9.1. Most of the unusual values can be explained by the data quality flag. At night, other reasons can influence the measurements. For analysis of integrated fluxes rejected data will need to be filled in. Obviously, investigations to infer process relationships should exclude both flagged data and the gap-filled values.

3.5 Site dependent quality control

Besides the quality classification of a single measurement series, classification of the site-specific data quality is needed to compare different sites within a network like FLUXNET for a better interpretation of experimental and modeled data. The data quality differs because of topography and this must be taken into account by comparison of the data quality. This quality check was developed to include footprint information (Foken et al. 2000). There are two different points of interest: The first point is the area of interest (e. g. a spruce forest) in the footprint of the measurements. The second points concerns the question: for which footprint areas can good data quality be assumed?

A program package has been developed (Göckede et al. 2003) and used for 18 CarboEurope eddy covariance measuring sites. The land use information of the surrounding is given by input matrices. Together with necessary meteorological input parameters, the main iteration loop of the program starts with a footprint calculation employing a user-defined start value for the roughness length z_0. The integrated Schmid (1997) model produces characteristic dimensions defining the two-dimensional horizontal extension of each so-called effect-level ring. Using these dimensions, which sketch a discrete version of the source weight function, it is possible to assign a weighting factor to each of the cells of the roughness matrix. A new roughness length z_0-final is calculated as the mean value of all the cells within the source area under consideration of the weighting factors. The iteration loop starts again with the improved value of z_0-final as the input value for the footprint routine. In the next step, the land use structure within the computed source area is analyzed. The weighting factors of the last source weighting function results are used to calculate the contribution of each type of land use (which can be up to 20, as defined by the user) to the total flux. Due to certain restrictions of the footprint model concerning the necessary input parameters, a portion of the input data set cannot be processed. Most of the time, these problems occur during stable stratification, when the computed source area grows to an extent that makes the numerical algorithms unstable. Finally figures like Figure 9.2 for the Weidenbrunnen/Waldstein site near Bayreuth/Germany (50°08'N, 11°52'E, 775 m a.s.l.), can be constructed that give a flux distribution over a four month measuring period that depends on the footprint. The color of the grid elements characterize the part of the area of interest to the flux. Such pictures can help find the best wind directions and the best positions of the tower to link the fluxes with the underlying surface.

To produce the overall performance of the flux data quality for a specific site, the results of all the footprint calculations are combined with the data quality assessment. The products of the procedure are two-dimensional matrices and graphs that form a combination of all the footprint analyses for the specific site. These matrices show, for example, the dominating data quality class for each of the grid cells (mean value) of the matrix surrounding the tower, in combination with its contribu-

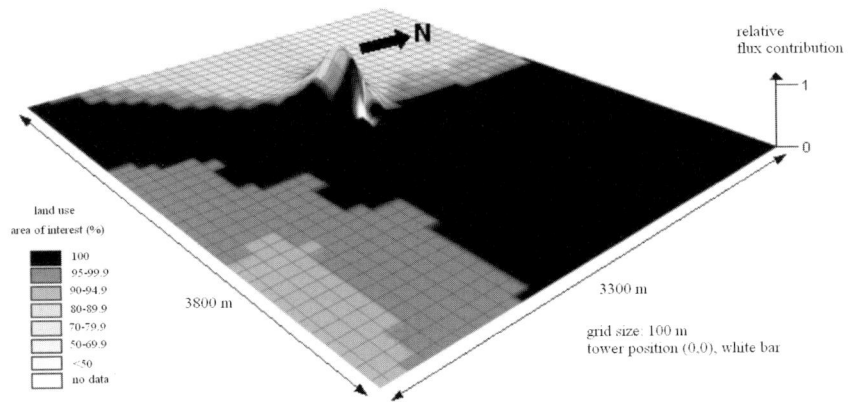

Figure 9.2. Quality analysis for the land use evaluation with flux contribution. Results were obtained with data from the Weidenbrunnen/Waldstein site for the period 01.05. – 31.08.1998 (Göckede et al. 2002).

tion to the total flux. This can be done for all types of fluxes. Only for scalar fluxes the quality flag of internal turbulence characteristics must be excluded in the near neutral case. As an example, the data quality distribution for the latent heat flux of Weidenbrunnen/Waldstein site is given in Figure 9.3. The lower data quality in western wind directions is caused by a clearing, which can also be indicated from the land use distribution (Figure 9.2). The low data quality in SWS direction (for stable stratification) is caused by the Waldstein mountain at a distance of 1.5 km. The possibility to bring data quality and possible influencing factors together is an application of the footprint model. Using the limit settings, the user of the program package can restrict the analysis to certain quality classes or a range of values for specific meteorological parameters, allowing a more detailed analysis under special conditions. The variation of these input parameters can also be performed automatically in a sequence mode with user defined upper and lower limits at specific increments.

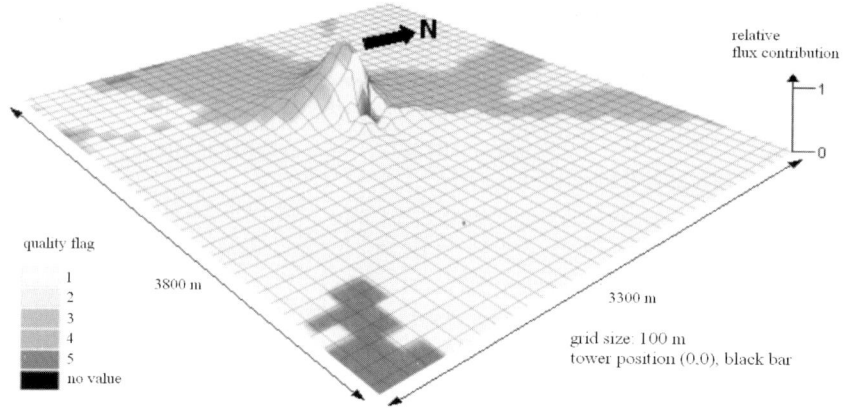

Figure 9.3. Quality flags for special distribution of the contribution to the latent heat flux. Results were obtained with data from the Weidenbrunnen/Waldstein site for the period 01.05.–31.08.1998 (Göckede et al. 2002).

4 Further Problems of Quality Control

Energy balance closure has often been used to identify the quality of eddy covariance measurements (Aubinet et al. 2000). For most of the sites a closure of the energy balance equation

$$R_n - H - \lambda E - G \pm \Delta S = \text{Res} \qquad (9.10)$$

with R_n net radiation, H sensible heat flux, λE latent heat flux, G ground heat flux, ΔS heat storage, is not zero but has a residual Res of approximately 10-20%. In some investigations of the energy balance closure problem (Culf et al. 2003, Foken and Oncley 1995, Oncley et al. 2002), the main reasons for this problem are errors of the sensors. For example the influence of net radiometers is significant because of the large part of net radiation in the energy balance. Measuring problems also exist of heat storage especially in the soil layer above the heat flux plates. Another reason is that mesoscale fluxes are not measured (Chapter 5. These reasons for the residual of the energy balance closure do not allow an energy balance closure as a correction factor for all turbulent fluxes or the use of energy balance closure as a measure of the data quality. However, there are many other studies where energy balance closure is consistently underestimated, without an identifiable cause. This has created some disparity among the methods employed by different groups. Some researchers use the energy balance closure as a further check, and adjust the CO_2 flux in the same proportion as the loss in the other turbulent fluxes (e. g., Amiro 2001, Barr et al. 2002). Some other researchers do not account for this turbulent loss, and consensus has not been reached in the research community. As an additional problem different instruments have different footprints.

The method of coordinate rotation also influences the data. Such rotations are necessary to align the x-axis with the mean wind (first rotation), to define a z axis so that the mean vertical wind component is zero (second rotation) and to rotate the system on the third axis so that the lateral momentum flux is zero (third rotation). This method was discussed by McMillen (1988) with a running mean as the reference coordinate system. Presently a rotation for each averaging interval (30 minutes) without the third rotation is proposed (Aubinet et al. 2000). This method is widely criticized because single events like convection, gusts, coherent structures etc., which have nothing to do with the coordinate system, are the reason for a significant rotation for a particular averaging interval. Even over low vegetation and flat terrain, rotation angles of 20-40° can be detected in the night and early morning hours. Therefore the planar-fit method (Wilczak et al. 2001) has been suggested (see Chapter 3) which rotates according the mean streamlines (Paw U et al. 2000). This streamline dependent coordinate system must be determined for one site and changes only with changes in the mounting of the sensor, with the time of the year (deciduous forest), with the wind speed (two classes) and the wind direction in heterogeneous and hilly terrain. The rotation angles are small and on the order of 2-5° and can be more with significant slope. After the rotation the data quality analysis as described in Sections 3.3 and 3.4 produces significant differences in the data quality especially for low wind velocities. As shown in Figure 9.4, the data quality is significantly lower for double rotation in comparison to planar fit in the classes of low friction velocity. The first method has low quality data typically for $u_* < 0.3$ m s^{-1} whereas the planar fit corresponds to approximately $u_* < 0.2$ m s^{-1}. This influence must be recognized, because it can influence the so-called u_*-criteria to correct nighttime carbon dioxide fluxes (Goulden et al., 1996).

Quality control procedures identify periods of unsuitable data, leaving non-random gaps in the dataset. The quality control procedures, instrument malfunctions, maintenance and calibration periods often remove 20 to 40% of the data. These gaps need to be filled for applications where long-term integrations are needed, though gaps should not be

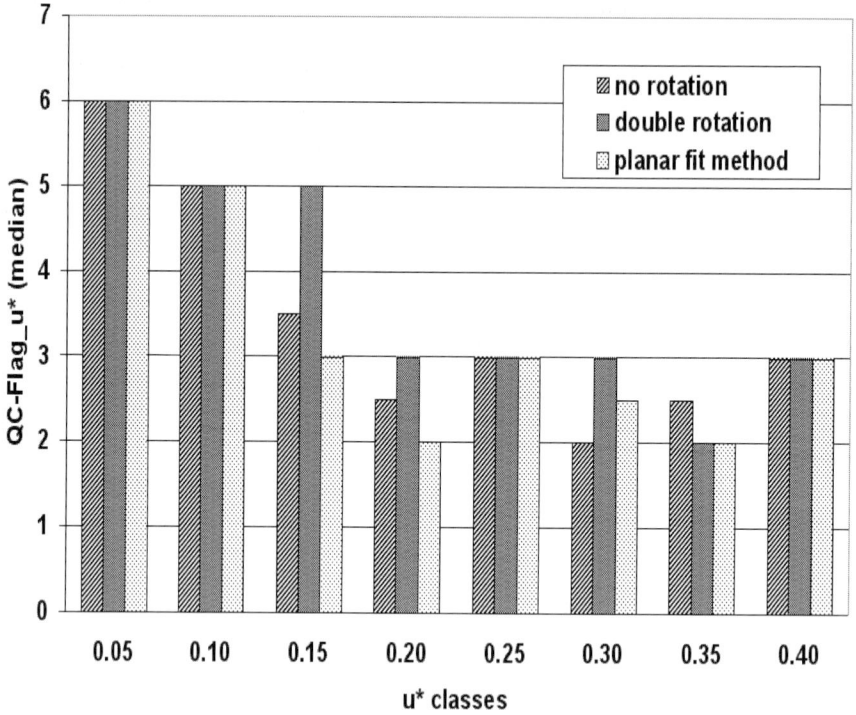

Figure 9.4. Data quality analysis for double rotation (Aubinet et al. 2000) and planar fit rotation (Wilczak et al. 2001) for measurements over an irrigated cotton field during EBEX-2000 (Oncley et al. 2002).

filled for process studies. Gap-filling creates additional uncertainty in the data, and there will always be a compromise between the use of possibly questionable flux data and replacement with values generated from a gap-filling algorithm. Confidence in gap-filling increases with knowledge and experience at any given flux site.

Falge et al. (2001a, 2001b) provide reviews of gap-filling strategies for energy flux and net ecosystem exchange (NEE) measurements. A variety of methods need to be applied, depending on the reasons for the gap creation. Nighttime gaps in NEE are best filled by either using soil respiration chambers or through developing a site-specific relationship between the respiration flux (mostly soil) and environmental variables such as soil temperature and moisture. Missing daytime NEE data can be estimated using physiological relationships that typically incorporate air temperature and light measurements. Short (e. g., a single half-hour period) gaps are usually filled through interpolation, whereas longer gaps may be estimated using the average of some period of good data for the same time of day. Gap-filling by averaging also needs to consider that gaps are often created by environmental conditions differing from the average, such as instrument malfunctions during heavy precipitation. The implications of gap-filling can be substantial, and in the case of NEE, can change the conclusions on the magnitude of annual carbon sequestration. Falge et al. (2001a) compared some gap-filling methods for NEE for 18 sites, illustrating that different methods could alter the annual sum of NEE by -45 to +200 $g\,C\,m^{-2}$, a significant portion of the total flux for some ecosystems. The conclusion is that quality-procedures need to focus on truly incorrect data since there is still a large uncertainty in filling gaps, and that the estimation of long-term fluxes can best be improved with good knowledge of the site processes.

Over a forest site the turbulence structure is very complicated (Amiro, 1990) sometimes with ramp structures mainly at daytime and wave structures (gravity waves) at nighttime (Chapter 8). The contribution of coherent structures to the whole flux is generally unknown. Well-organized ramp structures may be measured with the eddy covariance method. The determination of the flux due to ramp structures with the surface renewal method (Snyder et al., 1996) compares well with eddy covariance measurements (Rummel et al. 2002). In contrast, single coherent structures can indicate non-stationary conditions and be identified falsely as low quality data. We need continuously running procedures to calculate and control fluxes under these circumstances.

The decoupling of the atmosphere from the forest also needs to be considered. This is a typical situation during stable stratification at night. One must also consider the possibility of a mixing layer immediately above the forest canopy (Finnigan 2000, Raupach et al. 1996), caused by the high wind shear above the forest. The similarity analysis of the length scales of the shear layer and the coherent structures show that the forest and the atmosphere are often only coupled at daytime, often with strong coherent structures (Wichura et al. 2002).

One must also consider the mean transport at the upper boundary of a control volume (Chapter 10). Also the horizontal and vertical advective transport must be taken into account to interpret the vertical flux. An adequate choice of the coordinate system, for instance by planar fit rotation can help to interpret the vertical advection. Nevertheless, these site specific phenomena are difficult to check through automatic quality control procedures.

Plant physiological tests and ecosystem level measurements of carbon or water budgets can also be very useful in verifying the quality of the flux data. For example, soil chambers can give nighttime estimates of respiration during periods of weak turbulence when micrometeorological conditions fail. Plant leaf chambers can confirm the response of plants to certain conditions when turbulent flux measurements are questioned. Biomass inventories (Curtis et al. 2002) provide additional checks on annual integrals of flux data. The best possible estimate of net ecosystem exchange should combine a consistent set of independently determined quantities.

5 Conclusion

The quality assurance and quality control are outstanding problems that are incompletely fulfilled in most of the FLUXNET networks. For new stations a complete quality assurance plan can help provide a measuring system that can run within a short time on a high quality level. The quality control is always a combination of different levels of control and some very site-specific tests. Although an absolute uniform tool is impossible, a set of minimum standards is essential to ensure data comparability between sites in a network and over time for long-term measurements. Nevertheless some tools for electrical, meteorological and statistical tests are available. Not only the tests but the correction of the data are necessary to produce high quality data. Very important are tests on the fulfillment of the theoretical basis of the eddy covariance method as in the non-stationarity tests and the integral turbulence characteristic test. Important is the combination of all test results in an overall quality flag for the user of the data. A proposal is given in this Chapter, but only standardization makes flux measurements comparable. This Chapter included a footprint dependent quality analysis in the CarboEurope flux program. Such analysis helps to assess the data quality of different stations. Nevertheless, the data quality is only one part of the problem. Ecological reasons make stations with a lower quality important, if the investigated ecosystem does not allow better data qualities due to hilly terrain etc. The presented quality control tools work under most of the meteorological conditions especially over low vegetation. The measurement of nighttime fluxes, when the theoretical basis of the eddy covariance method fails, is not yet included in this procedure and the complicated turbulence structure over forests needs more investigation to find adequate algorithms to check the data.

Quality control and quality assurance tests are a fundamental part of the protocols used to arrive at good estimates of turbulent fluxes and

NEE. Many of the methods have been derived through experience by an ensemble of researchers. Although there is often a good reason for site-specific procedures, most of the scientific community has similar issues to address. Hence, networks are developing prescriptive procedures to achieve a basic level of data quality. Objective methods of removing spikes and identifying appropriate turbulent conditions, instrument malfunctions, non-stationary conditions, and appropriate fetch are common to all measurement sites. Decisions regarding coordinate rotation schemes, averaging periods, energy balance closure and gap-filling are less straightforward, and need to be further investigated to arrive at standard techniques. With the wide experience being gained through international FLUXNET collaborations, consensus on all of these procedures may be reached in the near future.

6 Acknowledgment

For developing the system of quality control procedures for eddy covariance measurements several scientists supported this work, mainly Corinna Rebmann, Christoph Thomas and Bodo Wichura from the University of Bayreuth, and Dean Vickers form the Oregon State University. The authors also thank Hans Peter Schmid (Indiana University) for using his footprint model. The work was supported by the European CARBOEUROFLUX and the INTAS program and the German afo-2000, VERTIKO and DEKLIM, EVA-GRIPS program, and the AMERIFLUX program.

7 References

Amiro, B. D.: 1990, 'Comparison of turbulence statistics within three boreal forest canopies', *Bound.-Layer Meteorol.* **51**, 99-121.

Amiro, B. D.: 2001, 'Paired-tower measurements of carbon and energy fluxes following disturbance in the boreal forest', *Global Change Biology* **7**, 253-268.

Arya, S. P.: 2001, *Introduction to Micrometeorology*, Academic Press, San Diego, 415 pp.

Aubinet, M., Grelle, A., Ibrom, A., Rannik, Ü., Moncrieff, J., Foken, T., Kowalski, A. S., Martin, P. H., Berbigier, P., Bernhofer, C., Clement, R., Elbers, J., Granier, A., Grünwald, T., Morgenstern, K., Pilegaard, K., Rebmann, C., Snijders, W., Valentini, R., Vesala, T.: 2000, 'Estimates of the annual net carbon and water exchange of forests: The EUROFLUX methodology', *Adv. Ecol. Res.* **30**, 113-175.

Barr, A. G., Griffis, T. J., Black, T. A., Lee, X., Staebler, R. M., Fuentes, J. D., Chen, Z., Morgenstern, K.: 2002, 'Comparing the carbon budgets of boreal and temperate deciduous forest stands', *Can. J. For. Res.* **32**, 813-822.

Betts, A. K., Desjardins, R. L., MacPherson, J. I., Kelly, R. D.: 1990, 'Boundary-layer heat and moisture budgets from FIFE', *Bound.-Layer Meteorol.* **50**, 109-138.

Beyrich, F., Herzog, H.-J., Neisser, J.: 2002, 'The LITFASS project of DWD and the LITFASS-98 Experiment: The project strategy and the experimental setup', *Theor. Appl. Climat.* **73**, 3-18.

Brock, F. V.: 1986, 'A nonlinear filter to remove impulse noise from meteorological data', *J. Atmos. Oceanic Techn.* **3**, 51-58.

Culf, A. D., Foken, T., Gash, J. H. C.: 2003, 'The energy balance closure problem', in: P. Kabat et al. (Editors), *Vegetation, Water, Humans and the Climate. A New Perspective on an Interactive System*, Springer, Berlin, Heidelberg, pp. 159-166.

Curtis, P. J., Hanson, P. J., Bolstad, P., Barford, C., Randolph, J. C., Schmid, H. P., Wilson, K. B.: 2002, 'Biometric and eddy-covariance based estimates of annual carbon storage in five eastern North American deciduous forests', *Agric. Forest Meteorol.* **113**, 3-19.

DeGaetano, A. T.: 1997, 'A quality-control routine for hourly wind observations", *J. Atmosph. Oceanic Techn.* **14**, 308-317.

Desjardins, R. L., Hart, R. L., MacPherson, J. I., Schuepp, P. H., Verma, S. B.: 1992, 'Aircraft- and tower-based fluxes of carbon dioxide, latent, and sensible heat', *J. Geophys. Res.* **97**, 18,477-18,485.

Desjardins, R. L., MacPherson, J. I., Mahrt, L., Schuepp, P., Pattey, E., Neumann, H., Baldocchi, D., Wofsy, S., Fitzjarrald, D., McCaughey, H., Joiner, D. W.: 1997, 'Scaling up flux measurements for the boreal forest using aircraft-tower combination', *J. Geophys. Res.* **102D**, 29125-29133.

Essenwanger, O. M.: 1969, 'Analytical procedures for the quality control of meteorological data', *Proc. Amer. Meteor. Soc. Symposium on Meteorological Observations and Instrumentation, Meteorol. Monogr.* **11(33)**, 141-147.

Falge, E., et al.: 2001a, 'Gap filling strategies for defensible annual sums of net ecosystem exchange', *Agric. Forest Meteorol.* **107**, 43-69.

Falge, E., et al.: 2001b, 'Gap filling strategies for long term energy flux data sets', *Agric. Forest Meteorol.* **107**, 71-77.

Finkelstein, P. L., Sims, P. F.: 2001, 'Sampling error in eddy correlation flux measurements', *J. Geophys. Res.* **D106**, 3503-3509.

Finnigan, J.: 2000, 'Turbulence in plant canopies', *Ann. Rev. Fluid Mech.* **32**, 519-571.

Finnigan, J. J., Clement, R., Malhi, Y., Leuning, R., Cleugh, H. A.: 2003, 'A re-evaluation of long-term flux measurement techniques, Part I: Averaging and coordinate rotation', *Bound.-Layer Meteor.* **107**, 1-48.

Foken, T.: 2003, *Angewandte Meteorologie, Mikrometeorologische Methoden*, Springer, Berlin, Heidelberg, 289 pp.

Foken, T., Jegede, O. O., Weisensee, U., Richter, S. H., Handorf, D., Gösdorf, U., Vogel, G., Schubert, U., Kirzel, H.-J., Thiermann, V.: 1997, 'Results of the LINEX-96/2 Experiment", *Deutscher Wetterdienst, Forschung und Entwicklung, Arbeitsergebnisse* **48** 75 pp.

Foken, T., Mangold, A., Rebmann, C., Wichura, B.: 2000, 'Characterization of a complex measuring site for flux measurements', *14th Symposium on Boundary Layer and Turbulence*, Am. Meteorol. Soc., Boston, 388-389.

Foken, T., Oncley, S. P.: 1995, 'Results of the workshop "Instrumental and methodical problems of land surface flux measurements"', *Bull. Am. Meteorol. Soc.* **76**, 1191-1193.

Foken, T., Skeib, G., Richter, S. H.: 1991, 'Dependence of the integral turbulence characteristics on the stability of stratification and their use for Doppler-Sodar measurements', *Z. Meteorol.* **41**, 311-315.

Foken, T., Wichura, B.: 1996, 'Tools for quality assessment of surface-based flux measurements', *Agric. Forest Meteorol.* **78**, 83-105.

Gilgen, H., Whitlock, C. H., Koch, F., Müller, G., Ohmura, A., Steiger, D., Wheeler, R.: 1994, 'Technical plan for BSRN data management', *WRMC, Techn. Rep.* **1**, 56 pp.

Göckede, M., Rebmann, C., Foken, T.: 2002, *Characterization of a Complex Measuring Site for Flux Measurements*, Univ. Bayreuth., Abt. Mikrometeorologie, Arbeitsergebnisse 20, 20 pp. [Available from University of Bayreuth, Dept. of Micrometeorology, D-95440 Bayreuth]

Göckede, M., Rebmann, C., Foken, T.: 2003, 'Use of footprint models for data quality control of eddy covariance measurements, *Agric. Forest Meteorol.* submitted.

Goulden, M. L., Munger, J. W., Fan, F.-M., Daube, B. C., Wofsy, S. C.: 1996, 'Measurements of carbon sequestration by long-term eddy covariance: method and critical evaluation of accuracy', *Global Change Biol.* **2**, 159-168.

Gurjanov, A. E., Zubkovskij, S. L., Fedorov, M. M.: 1984, 'Mnogokanalnaja avtomatizirovannaja sistema obrabotki signalov na baze', *EVM. Geod. Geophys. Veröff., R. II* **26**, 17-20.

Hall, C. D., Ashcroft, J., Wright, J. D.: 1991, 'The use of output from a numerical model to monitor the quality of marine surface observation', *Meteorol. Mag.* **120**, 137-149.

Handorf, D., Foken, T.: 1997, 'Analysis of turbulent structure over an Antarctic ice shelf by means of wavelet transformation', *12th Symosium on Boundary Layer and Turbulence*, Am. Meteorol. Soc., 245-246.

Haugen, D. A.: 1978, 'Effects of sampling rates and averaging periods on meteorological measurements', *Fourth Symp. on Meteorol. Observ. Instr.*, Am. Meteorol. Soc., 15-18.

Haugen, D. H. (Editor): 1973, *Workshop on Micrometeorology*, Am. Meteorol. Soc., 392 pp.

Hicks, B. B.: 1981, 'An examination of the turbulence statistics in the surface boundary layer', *Bound.-Layer Meteorol.* **21**, 389-402.

Højstrup, J.: 1993, 'A statistical data screening procedure', *Meas. Sci. Techn.* **4**, 153-157.

Howell, J. F., Sun, J.: 1999, 'Surface layer fluxes in stable conditions', *Bound.-Layer Meteorol.* **90**, 495-520.

Johansson, C., Smedman, A., Högström, U., Brasseur, J. G., Khanna, S.: 2001, 'Critical test of Monin-Obukhov similarity during convective conditions', *J. Atm. Sci.* **58**, 1549-1566.

Kaimal, J. C., Finnigan, J. J.: 1994, *Atmospheric Boundary Layer Flows: Their Structure and Measurement*, Oxford University Press, New York, NY, 289 pp.

Kaimal, J. C., Gaynor, J. E., Zimmerman, H. A., Zimmerman, G. A.: 1990, 'Minimizing flow distortion errors in a sonic anemometer', *Bound.-Layer Meteorol.*, **53**, 103-115.

Lenschow, D. H., Mann, J., Kristensen, L.: 1994, 'How long is long enough when measuring fluxes and other turbulence statistics?' *J. Atmosph. Oceanic Techn.* **11** 661-673.

Leuning, R., Judd, M. J.: 1996, 'The relative merits of open- and closed-path analyzers for measurements of eddy fluxes', *Global Change Biology*, **2**, 241-254.

Lorenc, A. C., Hammon, O.: 1988, 'Objective quality control of observations using Bayesian methods: theory, and a practical implementation', *Quart. J. Roy. Meteorol. Soc.* **114**, 515-544.

Lumley, J. L., Panofsky, H. A.: 1964, *The Structure of Atmospheric Turbulence*, Interscience Publishers, New York, 239 pp.

Mahrt, L.: 1991, 'Eddy asymmetry in the sheared heated boundary layer', *J. Atm. Sci.* **48**, 472-492.

Mahrt, L.: 1998, 'Flux sampling errors for aircraft and towers', *J. Atmosph. Oceanic Techn.* **15**, 416-429.

Mahrt, L., Moore, E., Vickers, D., Jensen, N. O.: 2001a, 'Dependence of turbulent and mesoscale velocity variances on scale and stability', *J. Appl. Meteorol.* **40**, 628-641.

Mahrt, L., Vickers, D., Sun, J.: 2002, 'Spatial variations of surface moisture flux from aircraft data', *Adv. Water Res.* **24**, 1133-1142.

Mahrt, L., Vickers, D., Sun, J., Jensen, N. O., Jørgensen, H.: 2001b, 'Determination of the surface drag coefficient', *Bound.-Layer Meteorol.* **99**, 249-276.

Mann, J., Lenschow, D. H.: 1994, 'Errors in airborne flux measurements', *J. Geophys. Res.* **99**, 14519-14526.

McMillen, R. T.: 1988, 'An eddy correlation technique with extended applicability to non-simple terrain', *Bound.-Layer Meteorol.* **43**, 231-245.

Moncrieff, J. B., Massheder, J. M., DeBruin, H., Elbers, J., Friborg, T., Heusinkveld, B., Kabat, P., Scott, S., Søgaard, H., Verhoef, A.: 1997, 'A system to measure surface fluxes of momentum, sensible heat, water vapor and carbon dioxide', *J. Hydrol.* **188-189**, 589-611.

Obukhov, A. M.: 1960, 'O strukture temperaturnogo polja i polja skorostej v uslovijach konvekcii', *Izv. AN SSSR*, ser. Geofiz., 1392-1396.

Oncley, S. P., Businger, J. A., Itsweire, E. C., Friehe, C. A., LaRue, J. C., Chang, S. S.: 1990, 'Surface layer profiles and turbulence measurements over uniform land under near-neutral conditions, *9th Symp. on Boundary Layer and Turbulence*,Am. Meteorol. Soc., 237-240.

Oncley, S. P., Foken, T., Vogt, R., Bernhofer, C., Kohsiek, W., Liu, H., Pitacco, A., Grantz, D., Ribeiro, L., Weidinger, T.: 2002, 'The energy balance experiment

EBEX-2000', *15th Symp. on Boundary Layer and Turbulence*, Am. Meteorol. Soc., 1-4.

Oncley, S. P., Friehe, C. A., Larue, J. C., Businger, J. A., Itsweire, E. C., Chang, S. S.: 1996, 'Surface-layer fluxes, profiles, and turbulence measurements over uniform terrain under near-neutral conditions', *J. Atm. Sci.* **53**, 1029-1054.

Panofsky, H. A., Dutton, J. A.: 1984, *Atmospheric Turbulence - Models and Methods for Engineering Applications*, John Wiley and Sons, New York, 397 pp.

Paw U, K. T., Baldocchi, D., Meyers, T. P., Wilson, K. B.: 2000, 'Correction of eddy covariance measurements incorporating both advective effects and density fluxes', *Bound.-Layer Meteorol.* **97**, 487-511.

Raupach, M. R., Finnigan, J. J., Brunet, Y.: 1996, 'Coherent eddies and turbulence in vegetation canopies: the mixing-layer analogy', *Bound.-Layer Meteorol.* **78**, 351-382.

Rummel, U., Ammann, C., Meixner, F. X.: 2002, 'Characterizing turbulent trace gas exchange above a dense tropical rain forest using wavelet and surface renewal analysis', *15th Symp. on Boundary Layer and Turbulence*, Am. Meteorol. Soc., 602-605.

Sakai, R., Fitzjarrald, D., Moore, K. E.: 2001, 'Importance of low-frequency contributions to eddy fluxes observed over rough surfaces', *J. Appl. Meteorol.* **40**, 2178-2192.

Schmid, H. P.: 1997, 'Experimental design for flux measurements: matching scales of observations and fluxes', *Agric. Forest Meteorol.* **87**, 179-200.

Shearman, R. J.: 1992, 'Quality assurance in the observation area of the Meteorological Office', *Meteorol. Mag.* **121**, 212-216.

Smith, S. R., Camp, J. P., Legler, D. M.: 1996, *Handbook of Quality Control, Procedures and Methods for Surface Meteorology Data*, Center for Ocean Atmospheric Prediction Studies, TOGA/COARE, Technical Report, 96-3: 60 pp. [Available from Florida State University, Tallahassee, FL, 32306-3041.]

Snyder, R. L., Spano, D., Paw U, K. T.: 1996, 'Surface renewal analysis for sensible and latent heat flux density', *Bound.-Layer Meteorol.* **77**, 249-266.

Stull, R. B.: 1988, *An Introduction to Boundary Layer Meteorology*, Kluwer Acad. Publ., Dordrecht, Boston, London, 666 pp.

Sun, J., Howell, J. F., Esbensen, S. K., Mahrt, L., Greb, C. M., Grossman, R., LeMone, M. A.: 1996, 'Scale dependence of air-sea fluxes over the Western Equatorial Pacific', *J. Atm. Sci.* **53**, 2997-3012.

Thomas, C., Foken, T.: 2002, 'Re-evaluation of integral turbulence characteristics and their parameterizations', *15th Conference on Turbulence and Boundary Layers*, Am. Meteorol. Soc., 129-132.

Trevino, G., Andreas, E. L.: 2000, 'Spectral analysis of nonstationary turbulence', *Bound.-Layer Meteorol.* **95**, 231-247.

Vickers, D., Mahrt, L.: 1997, 'Quality control and flux sampling problems for tower and aircraft data', *J. Atmosph. Oceanic Techn.* **14**, 512-526.

Vickers, D., Mahrt, L.: 2003, 'The cospectral gap and turbulent flux calculations', *J. Atmosph. Oceanic Techn.* in press.

Wichura, B., Buchmann, N., Foken, T.: 2002, 'Carbon dioxide exchange characteristics above a spruce forest, *25th Symp. Agric. Forest Meteor.*, Am. Meteorol. Soc., 63-64.

Wilczak, J. M., Oncley, S. P., Stage, S. A.: 2001, 'Sonic anemometer tilt correction algorithms', *Bound.-Layer Meteorol.* **99**, 127-150.

Wyngaard, J. C., Cot, O. R., Izumi, Y.: 1971, 'Local free convection, similarity and the budgets of shear stress and heat flux', *J. Atm. Sci.* **28**, 1171-1182.

Chapter 10

ADVECTION AND MODELING

John Finnigan
John.Finnigan@csiro.au

Abstract

Horizontal heterogeneity in either the source-sink distribution or the wind field results in streamwise advection of momentum and scalars, which must be accounted for whenever we deduce surface exchange from micrometeorological measurements on flux towers. This Chapter focuses on the second of these causes, addressing scalar advection in topography covered with uniform forest canopies rather than that generated by heterogeneity in the land cover. After defining advection and its relationship with modeling we discuss flow over forested hills by looking first at the wind field, next at the transfer of a generic scalar and finally at the implications for measuring photosynthesis on a two-dimensional ridge. Using analytic approaches as far as is possible, we show that both the turbulent wind field and scalar flow and transport in the canopy on a hill have a two-layer asymptotic structure with an upper canopy layer, coupled by turbulent transfer to the surface-layer flow above, and a lower canopy layer, that is driven by the pressure gradient produced as the wind field is deflected over the hill. The dynamics of these two layers are quite different and their matching through the upper canopy leads to strong modulation of turbulent transport over the hill and substantial advective flux divergence, even on gentle hills. The effect of the hill-induced perturbations on photosynthesis is calculated numerically and is shown to be small, being of order of the hill slope. In contrast, their effect on the net ecosystem exchange that would be deduced from eddy-flux measurements on a single flux tower is large, being of order one.

1 Introduction

For over forty years the lynch pin of micrometeorology has been the understanding of quasi-stationary boundary-layer flows over homogeneous terrain. Although studies of advection were among the earliest forays away from the heartland of 'flat earth' micrometeorology (e. g., Rider et al. 1963, Dyer and Crawford 1965, Bradley 1968) and measurements of flow over topography in both the wind tunnel and field had a brief flowering in the 1970s and 1980s (Kaimal and Finnigan 1994), until very recently inhomogeneous flows attracted quite a small part of the total effort in our field. Part of the reason for this was undoubtedly the considerably greater effort required to make field measurements in complex terrain with the need to deploy arrays of towers and to duplicate expensive instruments. The relative paucity of wind tunnel simulations is less easily explained.

Whether measurements are made in wind tunnels or in the open air, allowing inhomogeneity in two or there dimensions instead of just the vertical multiplies enormously the set of possible configurations that we would wish to investigate. Add to this the necessity to interpolate between what are always fewer measurement locations than are ideal and we can see why modeling has always played a bigger role in advection studies than in one-dimensional micrometeorology. Surveys of the field (see Kaimal and Finnigan 1994, Chapter 3 and references therein) record as many or more mathematical simulations of advective flows as experiments. Precisely because of the expense and difficulty of making measurements in inhomogeneous terrain, workers have turned to model studies from the outset to guide experimental design and to interpret the results.

In this Chapter we will concentrate on scalar advection generated by topography. This forms a small part of the totality of advection studies but is an area of particular relevance to the FLUXNET (Baldocchi et al. 2000). We will see that measurements are very scarce indeed in this domain and much of what we can say will be deductions from models. For this reason, we will rely, where we can, on analytic modeling approaches as these give the greatest insight into the underlying physics.

In the following sections we will discuss[1] in turn, modeling and its particular relevance to long term flux measurement in complex topography, and advection: what we mean by it and its relationship to the other aerodynamically important terms. In Section 3 we introduce an analytic model of the wind field over a two-dimensional ridge covered with a tall canopy and in Section 4 show how this wind field can be

---

[1] Our discussion uses the following notation: Vector and tensor quantities are denoted by bold face type, e. g., the velocity vector **u**, or, when appropriate, by a set of components, e. g., $\{u, v, w\}$. Standard meteorological notation is used and we employ right-handed rectangular Cartesian coordinates throughout. Averaging or filtering operators are denoted by an overbar and stochastic departures from the averaged or filtered variable by a prime thus, $c(t) = \bar{c} + c'(t)$. Other notation is introduced as encountered in the text.

used to drive first an analytic model of the transfer of a generic scalar and then a numerical model of photosynthesis on forested hills. The emphasis throughout will be on the consequences for flux measurement. Finally we will discuss the limitations of our models as well as their extension to calculate low frequency eddy fluxes. As we investigate the two topics of advection and modeling we will refer to subjects that are treated in detail elsewhere in this book. In particular we will rely on the discussions of averaging and filtering and of coordinate systems in Chapters 2 and 3.

2 General Remarks on Modeling and Advection

2.1 Modeling

In modeling the exchange of quantities between surface and atmosphere we have to address the action of the turbulent wind field, the radiation field, soil moisture dynamics and the physiology of the plants. Because here our focus is on advection we restrict our attention primarily to the role of the wind field. This does not imply that wind and turbulence exert the dominant controls on canopy-atmosphere transfer; rather that when we set out to use atmospheric measurements to infer this transfer, it is critical that we understand their behavior. The proximate aim of modeling is always to predict surface exchange in terms of primary quantities but we can usefully split the ultimate aims into three:

- The first of these is to frame and test hypotheses. For systems as complex as the terrestrial biosphere, representing key processes in models may be the simplest means of stating a set of hypotheses.

- The second aim is prediction. The surface-atmosphere interface provides the boundary condition for regional and global climate models so that capturing its dynamics in parameterizations and algorithms is a major component of atmospheric modeling at all scales from patch to global.

- The third aim takes a more operational stance. Aerodynamic approaches to measuring surface exchange proceed by data assimilation, whereby the measurements we can make are combined with a model of the processes we cannot measure to yield estimates of surface exchange compatible with both the measurements and the model. Such an approach is standard in many fields such as air quality forecasting and weather prediction, where meteorological models are both initialized and continuously adjusted by assimilated data (Kuo and Schlatter 1990).

Data assimilation methods have not yet been used explicitly to adjust long-term tower measurements for the effect of complex terrain. They are used implicitly, however, whenever we deduce surface exchange from measurements of the eddy flux and storage, the model into which the data are assimilated being a one-dimensional expression of the mass balance (Finnigan et al. 2003). In more complex applications the inverse modeling stage is always preceded by a 'forward modeling' phase where the ability of the candidate model to reproduce observations is tested. In this Chapter we shall concentrate on models of air flow over topography covered with tall canopies and the advective flux diversion that results. Although eventually we want to use these models as a basis for data assimilation, at present we are still in the forward-modeling phase and in many cases still trying to understand the key processes involved so the models we will discuss belong most properly in the first of the categories above: hypothesis testing.

Occupying a position between field measurements and mathematical modeling is physical modeling, usually in wind tunnels. Here we abstract some aspect of outdoor reality for closer examination. Dimensional analysis tells us which combinations of parameters must be kept constant, if we wish to reproduce full-scale dynamics in miniature. Inevitably we find that not all dimensionless groups can be kept constant as we shrink the length scale and hypotheses about the relative importance of different dynamical processes must then be made. The process of modeling and scaling is itself revealing and the detailed measurements possible at model scale are a powerful aid to mathematical parameterizations. We will see examples of the interplay between mathematical modeling, wind tunnel experiments and field measurements in the sections to follow.

2.2 Advection

The expression of the average scalar mass balance at a point in space is,

$$\frac{\overline{\partial c}}{\partial t} + \nabla.\overline{\mathbf{u}c} = \overline{S}(\mathbf{x})\delta(\mathbf{x} - \mathbf{x}_0) \tag{10.1}$$

where δ is the Dirac delta function and the overbar denotes an averaging or filtering operation that commutes with spatial differentiation. For the assumptions involved in this expression see Finnigan et al. (2003) and references therein. To use this expression as a basis for computing the surface exchange from eddy-flux measurements on towers we integrate it over a control volume V whose lower boundary is a representative patch of vegetated surface,

$$\int_V \frac{\overline{\partial c}}{\partial t} dV + \int_V \nabla \cdot \overline{\mathbf{u} c} \, dV = \int_V \overline{S}(\mathbf{x}) \delta(\mathbf{x} - \mathbf{x}_0) \, dV \qquad (10.2)$$

and dV is an elementary volume (see also Chapter 6). The form that the spatial and temporal derivative terms take is determined by both the coordinate system in which we choose to operate, and by the form of the averaging or filtering operator. As discussed by Lee et al. in Chapter 3, when the averaged wind field is used to define the vector basis of the coordinate system, these are not independent.

We see from Equations 10.1 and 10.2 that the mass balance equation contains the total scalar covariance, $\overline{\mathbf{u} c}$, not just the eddy covariance $\overline{\mathbf{u}' c'}$. Conventionally we split the total covariance into turbulent and mean fluxes and the simplest way to effect this split is via a simple time average,

$$\overline{c(t)} = \overline{c} = \frac{1}{T} \int_0^T c(t) dt; \quad c(t) = \overline{c} + c'(t); \quad \mathbf{u}(t) = \overline{\mathbf{u}} + \mathbf{u}'(t) \qquad (10.3)$$

from which it follows that

$$\overline{\mathbf{u} c} = \overline{\mathbf{u}} \, \overline{c} + \overline{\mathbf{u}' c'} \qquad (10.4)$$

Alternatively we can use a temporal filter,

$$\overline{c}(t) = \int_{-T_a}^{T_a} G(s-t) c(s) ds; \quad c(t) = \overline{c} + c'(t); \quad \mathbf{u}(t) = \overline{\mathbf{u}} + \mathbf{u}'(t) \qquad (10.5)$$

where $G(t)$ is the filter shape with a width $2T_a$. If $G(t)$ is defined as,

$$G(t) = \begin{cases} 1/2T_a & |t| \leq T_a \\ 0 & |t| > T_a \end{cases} \qquad (10.6)$$

then the filter is a simple moving average operation. Note that for the purposes of this Chapter we can regard detrending as a filtering operation.

Now we find that,

$$\overline{\mathbf{u} c} = \overline{\mathbf{u}} \, \overline{c} + \overline{\overline{\mathbf{u}} c'} + \overline{\mathbf{u}' \overline{c}} + \overline{\mathbf{u}' c'} \qquad (10.7)$$

So time-averaged covariances can be decomposed into means and fluctuations according to Reynolds averaging rules (Equation 10.4) whilst filtered or detrended covariances acquire the extra 'Leonard terms', the second and third terms on the RHS of Equation 10.7.

Several considerations motivate the split into means and fluctuations. A historical reason was the propensity of fast response sensors to have serious zero drift so that the turbulent fluctuations and the means had to be measured with different instruments. This motivation has been largely removed by sensor improvements. A second reason that is still important is the need to use the mean wind field to define coordinate directions as described in Lee et al. in Chapter 3 and in Finnigan (2004). A more elusive but still pervasive motivation for many workers is the formal split between ensemble-averaged and turbulent quantities that is employed in most textbooks on turbulence to derive the conservation equations. It seems an intuitive step to separate the fast and slowly varying parts of the key variables this way. Unfortunately it is impractical to actually employ ensemble averages in field experiments. The best approximation to ensemble averaging is the use of temporal filters (Equation 10.6; Chapter 8). A fourth motivation is that we sometimes want to separate means and fluctuations to apply data quality control measures that are based upon expectations of turbulent behavior in statistically stationary flows; we expect that the turbulent part of the signal will approximate this behavior (Chapter 9).

The final reason for a split of turbulent quantities into means and fluctuations is that mathematically and conceptually we treat them differently in mathematical models. We represent the mean fields as if they were deterministic quantities while turbulent components are handled statistically. While this conceptual split has a rigorous foundation in ensemble averaging, it is not easily reflected using time averages or filters unless there exists a 'spectral gap', separating atmospheric motions on longer scales from the turbulent eddies in the boundary layer that are actively transporting momentum and scalars. If we can place the averaging period T in this spectral gap, then the turbulent motions will be varying on much shorter time scales than the means and on time scales that characterize changes in the means, the turbulence moments may approximate statistical stationarity.

Such a spectral gap in the near-surface wind spectrum was recorded by Van der Hoven (1957) and has been reproduced in several textbooks since then so that the concept of the spectral gap is widely accepted. However, recent measurements by Ayotte et al. (2001) of horizontal wind speeds at 40 m from an array of towers show no such gap (Figure 10.1). These spectra, which comprise 8 years of data from sensitive cup anemometers on 12 towers, show no spectral gap in the horizontal wind and a steady increase in the spectral density of horizontal wind variance between the lowest resolved period of approximately 8 minutes and the Rossby wave period of approximately 3-4 days. Interestingly, the spectral shapes cor-

Advection and Modeling

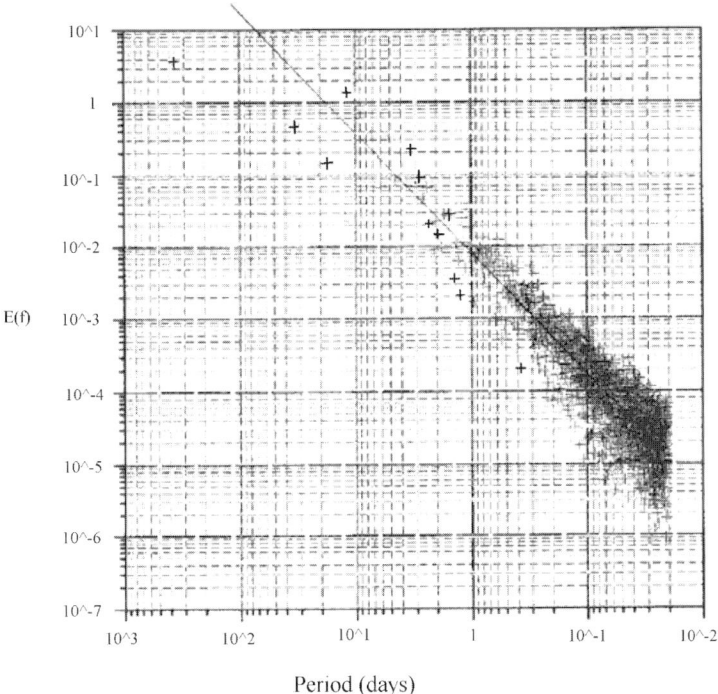

Figure 10.1. Power spectrum of horizontal wind speed measured at 40-m. Measurements were made on an array of towers in NSW, Australia. Figure reproduced with kind permission of Ayotte et al. (2001).

respond to those collated from measurements in the upper troposphere by Nastrom and Gage (1985), indicating that longer period variations in surface-layer horizontal wind speed may be due to tropospheric forcing of the planetary boundary layer (PBL) rather than instabilities within it.

In Section 4 below we will describe mechanisms whereby horizontal variations in wind speed can generate variations in scalar concentration correlated with variations in vertical velocity even when the scalar source or sink is uniform. Hence these long period variations in horizontal wind can translate into low frequency eddy fluxes. Finnigan et al. (2003) show that at some tall canopy sites the averaging period T, necessary to capture all significant contributions to eddy flux is approximately 4 hours, much longer than the conventionally accepted position of the spectral gap and longer than most of the eddy motions that can be generated within the PBL (see also Chapters 2 and 5).

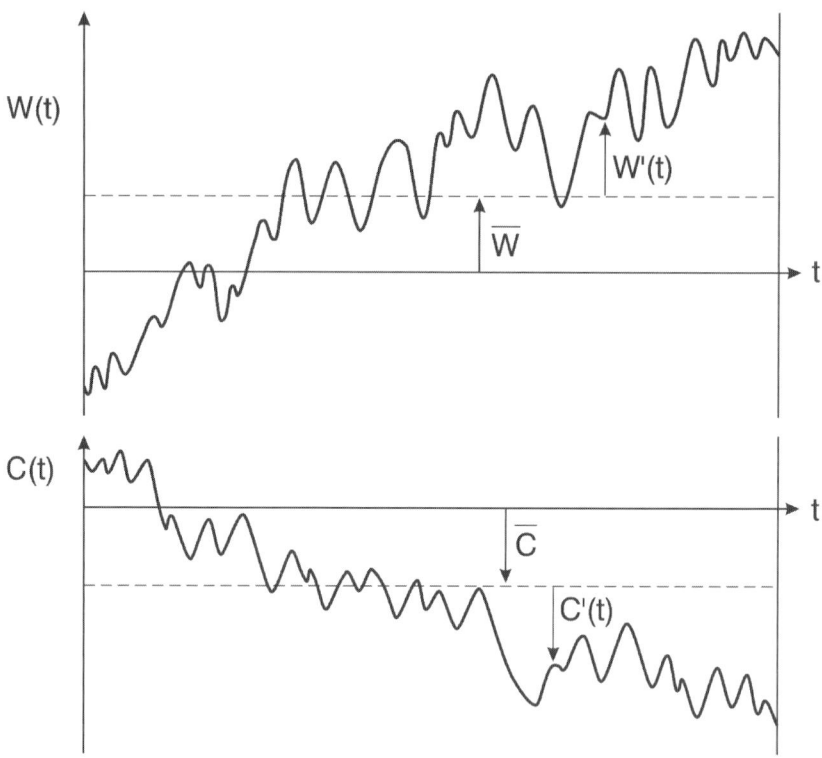

Figure 10.2. The effect of a block time average on a turbulent signal with a low frequency 'trend'.

When no spectral gap exists, the choice of period separating means and fluctuations can be arbitrary although Moncrieff et al. in Chapter 2 give a rational procedure for choosing such a period. Ideally the separation should be done by a filter (Equation 10.6) because with an appropriate choice for filter function the 'leakage' between means and fluctuations represented by the Leonard fluxes in Equation 10.7 can be made small, whereas if we use a time average (Equation 10.3), any slow variation of the flow is counted as turbulence (Figure 10.2) with consequences for model parameterizations as we shall see later. More work needs to be done to resolve properly the question of separating means and fluctuations. Nevertheless, almost all model development (and most data analysis currently) assumes that the mean fields are approximately steady and can be separated from the turbulence by a time average.

Advection and Modeling

If we express Equation 10.1 in Cartesian coordinates and apply the decomposition (Equation 10.4) we obtain,

$$\underbrace{\frac{\partial \overline{c}}{\partial t}}_{1} + \underbrace{\overline{u}\frac{\partial \overline{c}}{\partial x}}_{2} + \underbrace{\overline{v}\frac{\partial \overline{c}}{\partial y}}_{3} + \underbrace{\overline{w}\frac{\partial \overline{c}}{\partial z}}_{4} + \underbrace{\frac{\partial \overline{u'c'}}{\partial x}}_{5} + \underbrace{\frac{\partial \overline{v'c'}}{\partial y}}_{6} + \underbrace{\frac{\partial \overline{w'c'}}{\partial z}}_{7} = \underbrace{\overline{S}(\mathbf{x})\delta(\mathbf{x} - \mathbf{x}_0)}_{8}$$
(10.8)

where we have used the continuity condition,

$$\nabla.\overline{\mathbf{u}} = \nabla.\mathbf{u}' = 0$$

to write the mean or *advective* flux divergence, $\nabla.\overline{\mathbf{u}}\,\overline{c}$ as $\overline{\mathbf{u}}.\nabla\overline{c}$. When the wind field is horizontally homogeneous and steady and the source term, 8 is horizontally uniform, the advective terms 2, 3 and 4 of Equation 10.8, together with the terms 5 and 6, that express the horizontal divergence of eddy flux, are identically zero. Equation 10.8 then reduces to the familiar one-dimensional form that is commonly used by default to analyze flux tower data,

$$\frac{\partial \overline{c}}{\partial t} + \frac{\partial \overline{w'c'}}{\partial z} = \overline{S}(\mathbf{x})\delta(\mathbf{x} - \mathbf{x}_0)$$
(10.9)

As soon as horizontal variability[2] is introduced into the source term 8, however, we can expect terms 2-6 to be non-zero so that Equation 10.9 is no longer valid. The effects of simple changes in the scalar source, such as occur at the edge of a field or forest, have been studied and modeled for many years. See for example Kaimal and Finnigan (1994) and the references contained there. In models of that situation attention has been focused primarily on the changes to the wind field that accompany changes in surface roughness or energy balance or on changes in the scalar transfer calculated under the assumption that the wind field has not changed at the transition. In both cases the changes to the wind and scalar fields are confined to internal boundary layers that grow downwind of the surface transition. There are few studies that model the combined effect of changes to the wind field and to the scalar source, although this of course is the usual situation in practice. Furthermore, most model studies treat the situation where the changes in source/sink strength or roughness are confined to the surface plane but when the transition is

---

[2]We are assuming that the point-to-point variation in properties that occurs in the canopy airspace has been removed by the volume average operation implicitly whenever canopy processes such as sources and sinks are represented as smooth functions such as $\overline{S}$ in Equations 10.8 or 10.9 (see Finnigan 2000 and references therein). Hence, the horizontal variability we refer to here is taken to be on a scale much larger than leaf or plant spacing.

from a field to a forest or an urban area (urban canopy), we are interested in resolving the spatial structure of the wind and scalar fields through the height of the canopy. Thus far, this situation seems to have been addressed only for momentum transfer (Finnigan 2002, Belcher et al. 2003).

The second important source of advection is the wind field. Horizontal variation in the wind field can produce substantial advection even when the scalar source term $\overline{S}$ is uniform. The most obvious situation where this occurs is when we have a uniform scalar source such as a plant canopy on topography. Raupach et al. (1992) (henceforth RWCH) analyzed scalar transfer from the surface of rough low hills and showed that there were three distinct causes of the advection: convergence-divergence of the mean flow streamlines; changes in scalar eddy fluxes; and changes in the surface shear stress. The treatment of RWCH was an important step forward, not least because it was an analytic model and so revealed the underlying biophysics of the processes in a way that numerical simulations never do. However, it did not treat the case of a tall plant canopy on topography, which is the situation of many FLUXNET sites. We will discuss the necessary extensions of RWCH to accommodate tall canopies in Section 4 below.

In general we have to deal both with topography and with horizontal changes in the scalar sources. For the rest of this Chapter, however, we will confine our attention to uniform canopies on topography for the following reasons: a) most FLUXNET sites are deliberately located in areas of uniform forest cover; b) the idea that advection can be forced by topography with a uniform surface source is less familiar than the case of a non-uniform surface source; c) recent model studies show that systematic bias in estimates of net ecosystem exchange (NEE) can occur if topographically forced advection is neglected (see Section 4.2 below); d) there is some evidence from the continental USA that the strongest terrestrial sinks of CO_2 coincide with montane geography (Schimel et al. 2003). Finally and perhaps most importantly, the analysis necessary to understand the effects of hills on canopy flow and transport reveals some important information about canopy dynamics that is of application to disturbed canopy flows in general.

3 The Turbulent Wind Field in a Tall Canopy on a Low Hill

The basic physics of neutrally stratified flow over hills were elucidated by Hunt, Liebovich and Richards (1988) (henceforth HLR), building on earlier work by Jackson and Hunt (1975) and others. In order to

Advection and Modeling

obtain the explanatory power of an analytic solution they restricted their attention to low hills. Despite this, the insights they obtained can be applied to general topography. Finnigan and Belcher (2004) (henceforth FB) have extended the model of HLR, replacing the rough surface of the HLR model with a plant canopy and their approach also yields an analytic solution in the limit of a 'tall canopy'.

In the following section we introduce some new notation: $U_B(z)$ is the undisturbed upwind flow (or the areally averaged flow in a region of continuous topography) while $\Delta \bar{p}$, $\Delta \tau(x,z)$ and $\Delta \bar{u}(x,z)$ are the perturbations to the mean pressure, shear stress and streamwise velocity respectively caused by the hill. Hence,

$$\bar{u} = U_B(z) + \Delta \bar{u}(x,z); \quad \bar{w} = \Delta \bar{w}(x,z);$$
$$\bar{p} = P_B + \Delta \bar{p}(x,z); \quad \tau = T_B(z) + \Delta \tau(x,z) \quad (10.10)$$

We also define the parameters, H, the height of the hill, L its horizontal length scale (distance from crest to half-height point) and h_c the canopy height. C_d is the dimensionless aerodynamic drag coefficient of the canopy foliage[3] and $a(z)$ is the foliage area per unit volume of space. The momentum absorption distance is defined as $L_c = 1/(C_d a)$.

The FB model considers the perturbations caused by the hill to a background flow that consists of a logarithmic profile above the canopy and an exponential profile within the canopy. These profiles are solutions to the one-dimensional momentum equations with a constant mixing length within the canopy and a mixing length proportional to $z + d$ above the canopy, where d is the displacement height of the logarithmic profile. Matching these two solutions at the top of the canopy, which is taken as $z = 0$, the origin of the vertical coordinate, we find,

$$U_B(z) = \begin{cases} \dfrac{u_*}{k} \ln\left(\dfrac{z+d}{z_0}\right) & \text{for } z > 0 \\[1em] U_h \exp\left(\dfrac{\beta z}{l}\right) & \text{for } z \leq 0 \end{cases} \quad (10.11)$$

where k is the von Karman constant, z_0 is surface roughness, $U_h = U_B(0)$ is the mean wind speed at the top of the canopy, u_* is the friction velocity, $l = 2\beta^3 L_c$ is the mixing length in the canopy, $l = ku_*(z-d)$ is the mixing length above the canopy and $\beta = u_*/U_h$ quantifies the mass flux through the canopy. For closed uniform natural canopies, $\beta \approx 0.3$

---

[3]In this Chapter, C_d is defined by the expression $F_D = C_D a \bar{u} |\bar{u}|$, where F_D is the aerodynamic drag and so differs by a factor of 2 from some definitions.

(Raupach et al. 1996). Matching both mean wind and shear stress at the canopy top also fixes the following relationships,

$$U_h = \frac{u_*}{k} \ln(\frac{d}{z_0}); \quad d = l/k; \quad z_0 = \frac{l}{k} \exp(-k/\beta) \quad (10.12)$$

The analyses of HLR and FB divide the flow in the canopy and in the free boundary layer above into a series of layers with essentially different dynamics. The dominant terms in the momentum balance in each layer are determined by a scale analysis and the eventual solution to the flow field is achieved by asymptotically matching solutions for the flow in each layer. The model applies in the limit that $H/L << 1$. By adopting this limit, HLR were able to make the important simplification of calculating the leading order perturbation to the pressure field using potential flow theory. This perturbation to the mean pressure, $\Delta \bar{p}(x, z)$ can then be taken to drive the leading order (i. e. $O[H/L]$) velocity and shear stress perturbations over the hill. Higher order corrections to the leading order terms can then be obtained using standard methods of perturbation analysis. See for example Van Dyke (1978).

HLR also showed that $\Delta \bar{p} \sim O[U_0^2 H/L]$, where $U_0 = U_B(h_m)$ is the streamwise velocity well above the surface (more precisely at the 'middle layer' height h_m defined below). Furthermore, they showed that $\partial \Delta \bar{p}/\partial x, \partial \Delta \bar{p}/\partial z \sim O[U_0^2 H/L^2]$. These deductions lead to important simplifications that permit an analytic solution to the flow field. The restriction to 'tall' canopies in the theory of FB can be interpreted as a requirement that almost all the momentum flux is absorbed as aerodynamic drag on the foliage and not as shear stress on the underlying surface.

In the free boundary layer above the canopy the flow divides asymptotically into an outer layer, where the flow perturbations caused by the hill are essentially an inviscid response to the pressure forcing, and an inner or shear stress layer where changes to the shear stress caused by the hill play a role at first order in the momentum balance. The momentum balance in the outer layer to $O[H/L]$ becomes,

$$\text{Outer layer} \quad U_B \frac{\partial \Delta \bar{u}}{\partial x} + \Delta \bar{w} \frac{\partial U_B}{\partial z} = -\frac{\partial \Delta \bar{p}}{\partial x} \quad (10.13)$$

and the middle layer height h_m, which divides the outer layer into a lower part, where vorticity in the background flow is dynamically important, and an upper part with potential flow, is defined by the relationship,

$$\frac{h_m}{L} \ln^{1/2}(h_m/z_0) = 1 \quad (10.14)$$

Advection and Modeling

In the shear stress layer of depth h_i the perturbation in shear stress divergence becomes important so the streamwise momentum balance is,

$$\text{Shear stress layer (of depth } h_i \text{)} \quad U_B \frac{\partial \Delta \bar{u}}{\partial x} + \Delta \bar{w} \frac{\partial U_B}{\partial z} = -\frac{\partial \Delta \bar{p}}{\partial x} + \frac{\partial \Delta \tau}{\partial z} \tag{10.15}$$

The depth of the shear stress layer is defined by the implicit relation,

$$\frac{h_i}{L} \ln(h_i/z_0) = 2k^2 \tag{10.16}$$

The canopy flow itself breaks down into upper and a lower canopy layers. In the upper canopy the linearized momentum balance to $O[H/L]$ is,

$$\text{Upper canopy layer} \quad 0 = -\frac{\partial \Delta \bar{p}}{\partial x} + \frac{\partial \Delta \tau}{\partial z} - \frac{2 U_B \Delta \bar{u}}{L_C} \tag{10.17}$$

and we see that in the canopy advection is small compared to the retained terms. As we get deeper into the canopy, the shear stress gradient, $\partial \Delta \tau / \partial z$ becomes weaker and the lower canopy flow reduces to a balance between the pressure gradient and the drag,

$$\text{Lower canopy layer} \quad 0 = -\frac{\partial \Delta \bar{p}}{\partial x} - \frac{1}{L_C}(U_B + \Delta \bar{u})|U_B + \Delta \bar{u}| \tag{10.18}$$

The form of the aerodynamic drag term follows because the drag force always opposes the velocity and we see also that in the lower canopy the momentum balance is non-linear. This is because the background velocity U_B decays exponentially into the canopy (Equation 10.11), while the driving pressure gradient $\partial \Delta \bar{p}/\partial x$, which varies on a length scale L in the vertical, penetrates the canopy relatively undiminished so that in the lower canopy it must be balanced primarily by the perturbation drag force, $L_C^{-1} \Delta \bar{u} |\Delta \bar{u}|$.

Equation 10.18 is, consequently an algebraic equation with the solution,

$$\text{Lower canopy layer} \quad \Delta \bar{u}(x) = -\sqrt{L_C^{-1} |\partial \Delta \bar{p}/\partial x|} \operatorname{sgn}(\partial \Delta \bar{p}/\partial x) \tag{10.19}$$

We can see from Equation 10.19 that the largest velocity perturbations in the lower canopy will coincide with the position of the largest negative pressure gradient over the hill, which is well upwind of the hill crest.

We can also see from Equation 10.13 that to first order, the velocity perturbations in the outer layer will vary as the square root of (minus) the pressure perturbation, that is,

$$\text{Outer layer} \qquad \Delta\bar{u}(x,z) \propto -\sqrt{|\Delta\bar{p}|}\,\text{sgn}(\Delta\bar{p}) \qquad (10.20)$$

As we travel over a low hill, the pressure falls, reaches a minimum just before the crest and then rises again in the lee. The minimum pressure, therefore, occurs almost at the crest and the minimum pressure gradient occurs about half way up the windward slope of the hill. From Equations 10.19 and 10.20 we can see that the perturbation in the outer layer flow peaks roughly over the hill crest and the deep-canopy flow perturbation peaks on the upward face of the hill and is passing through zero on the hill crest, becoming negative on the lee side.

The solutions of Equations 10.15 and 10.17 for the shear stress layer and the upper canopy layer are more complicated as we now have to account for shear stress divergence but we can note the following points. Because the shear stress layer is a region of local equilibrium (turbulence production $\sim$ dissipation) it is feasible to model the shear stress with a mixing length model,

$$\text{Shear stress layer} \qquad \tau = -\left[l^2 \frac{\partial \bar{u}}{\partial z}\right] \frac{\partial \bar{u}}{\partial z} \quad \text{with} \quad l = k(z-d) \qquad (10.21)$$

Within the canopy, surprisingly, it is also feasible to use a mixing length model for the perturbations although not for the background flow, hence the use of a mixing length for the background flow solutions (Equation 10.11) must be regarded as merely a convenient heuristic (FB). Within the canopy, in keeping with the 'mixing layer analogy' for canopy turbulence (Finnigan 2000), the dynamically correct mixing length is a constant. FB show that the shear stress in the upper canopy can be written,

$$\text{Upper canopy} \qquad \tau = -\left[l^2 \frac{\partial \bar{u}}{\partial z}\right] \frac{\partial \bar{u}}{\partial z} \quad \text{with} \quad l = 2\beta^3 L_C \qquad (10.22)$$

The matched solutions to the four layers tell us that the shear stress layer and upper canopy layers form a region of adjustment across which the mean flow perturbations change from being in phase with (minus) the pressure well above the surface (Equation 10.20) to being in phase with (minus) the pressure gradient deep in the canopy (Equation 10.19). This adjustment strongly modulates the shear across the canopy top. These features are clearly illustrated in Figure 10.3 from FB, where the streamwise velocity perturbations at a series of stations across one of

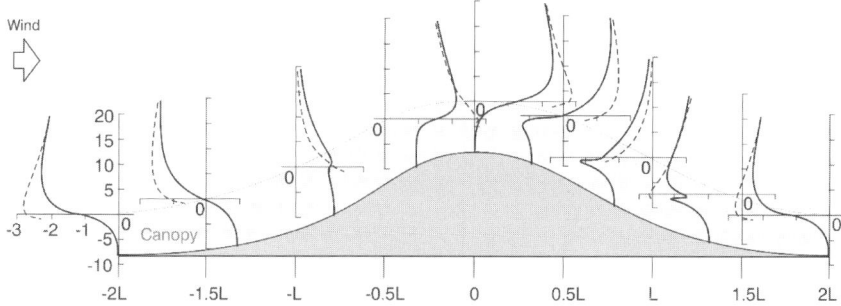

Figure 10.3. Comparison of dimensionless canopy velocity perturbation from the theory of FB (solid line) with the no-canopy HLR solution (dotted line). Variable plotted is $\Delta \bar{u}/U_{SC}$, where $U_{SC} = \dfrac{H}{L} U_0$. Note the HLR solution is only valid to $z = -d + z_0$. Profiles are plotted at a series of X/L values between $X/L = -2$ (upwind trough) and $X/L = 2$ (downwind trough) on one of a series of sinusoidal ridges. The units of Z are m and the vertical range is from $2h_i > Z > L_C$.

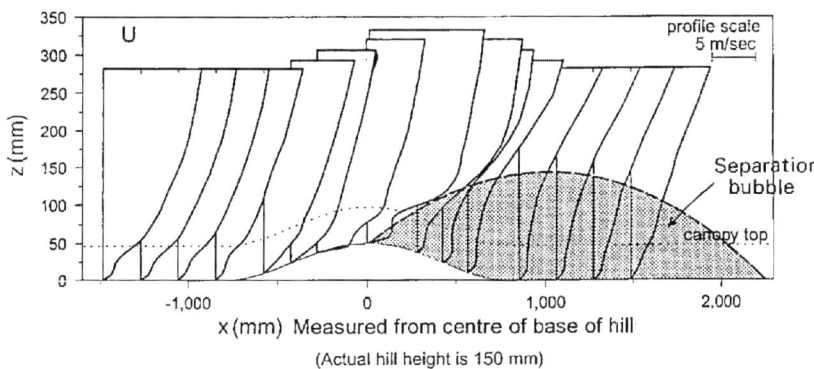

Figure 10.4. Mean velocity profiles on a wind-tunnel model study of flow over a steep two-dimensional ridge covered with a plant canopy. Figure reproduced from Finnigan and Brunet (1995).

a range of sinusoidal ridges are plotted. Included for comparison are solutions for a rough surface with the same z_0 from the theory of HLR. We can see that the extra turbulent mixing generated by the canopy reduces the sharp speed-up peak on the hill crest predicted by HLR and moves it from around $z \simeq h_i/3$ to $z \simeq h_i$.

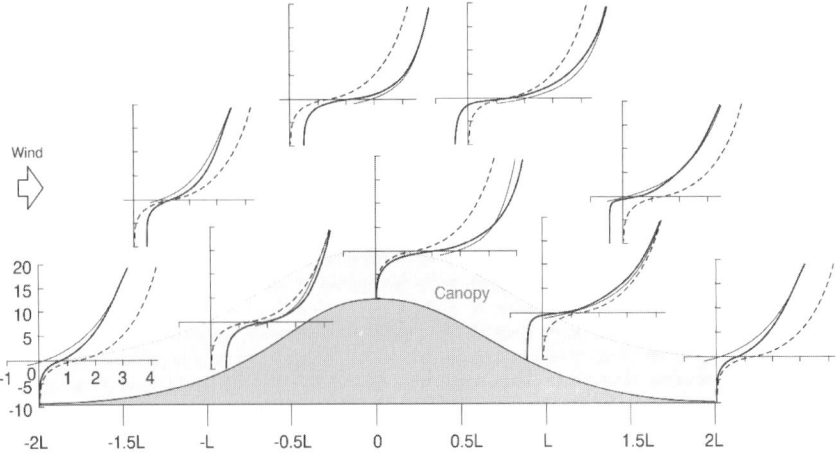

Figure 10.5. Comparison of dimensionless total velocity, $U_B + \Delta \bar{u}$ in the canopy (heavy solid line) with the no-canopy HLR solution (thin solid line) from the theory of FB. The background velocity $U_B(Z)$ is shown as a dashed line. Note the HLR solution is only valid to $z = -d + z_0$. Profiles are plotted at a series of X/L values between $X/L = -2$ (upwind trough) and $X/L = 2$ (downwind trough). Profiles at fractional X/L values are displaced upwards for clarity. The units of Z are m and the vertical range is from $2h_i > Z > L_c$.

In Figure 10.4 we show consecutive vertical profiles from the wind tunnel model study of Finnigan and Brunet (1995). Although this hill is too steep to satisfy the $H/L << 1$ limits of linear theory, upwind of the hill crest we can still see the main features predicted by the FB model. The maximum velocity in the lower canopy occurs well before the crest and is falling by the hilltop. The difference between lower canopy and outer layer velocities is a maximum at the hilltop and maximizes the canopy top shear at that point with consequences for the magnitude and scale of turbulence production. Conversely, the difference is at a minimum halfway up the hill, where the lower-canopy velocity is maximal but the outer layer flow has not yet increased much. This effect is so marked that the inflexion point in the velocity profile at the top of the canopy has disappeared. Note also that on this steep hill we observe a large separation bubble behind the hill crest.

One final prediction made by the theory of FB must be mentioned as it turns out to have important consequences for scalar transport. This is that even on hills of low slope ($H/L << 1$), a region of reversed flow will appear within the canopy on the lee side of the hill, if the canopy

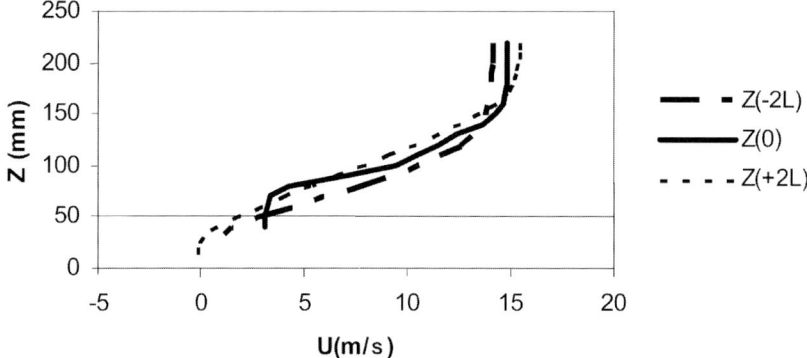

Figure 10.6. Laser-doppler anemometer profiles of mean velocity in and above a wind tunnel model study of flow over an isolated sinusoidal two-dimensional ridge covered with a tall canopy. Profiles are located upwind ($x = -2L$, dot dash line); on the hilltop (solid line) and downwind ($x = +2L$, dotted line) The hill parameters were $H/L = 0.1$, $h_c/H = 1.0$. Full details of the experiment can be obtained from the author (J. J. Finnigan, D. Hughes pers. comm.).

is sufficiently deep and dense. This is illustrated in Figure 10.5, where we have plotted the total velocity over the same hill and canopy as in Figure 10.3. As noted above, the driving pressure gradient, $\partial \Delta \bar{p}/\partial x$ passes essentially undiminished through the canopy and will produce a velocity perturbation $\Delta \bar{u}$ that is negative behind the hill crest where $\partial \Delta \bar{p}/\partial x > 0$ (Equation 10.19). Sufficiently deep into the canopy we can expect that $|\Delta \bar{u}| > U_B$ because $U_B(z)$ decays exponentially into the canopy. The condition for this region of flow reversal to appear (FB) is,

$$h_c > \frac{l}{2\beta} \ln \left[\frac{U_0^2}{U_h^2} (H/2) k^2 L_c \right] \qquad (10.23)$$

The appearance of this reversed flow region in a deep canopy on a shallow hill has now been confirmed by a wind-tunnel model study (J. J. Finnigan, D. Hughes pers. comm.). Figure 10.6 shows profiles of total velocity $U_B + \Delta \bar{u}$ at three x stations ($x = -2L; 0; 2L$) over the model hill. The hill parameters were $H/L = 0.1; h_c/H = 1.0$. The measurements were taken with a laser-doppler anemometer so that the reversed flow region could be properly resolved. Full details of the experiment can be obtained from the author. In the light of the above discussion, it is interesting to contrast this experiment with that of Finnigan and

Brunet (1995). Finnigan and Brunet conducted their experiment in a large boundary layer wind tunnel (working section 1m × 2m × 12m) with a carefully simulated atmospheric boundary layer as they aimed to model the full set of processes occurring in neutral flow over a forested hill. The experiment of Finnigan and Hughes in contrast took place in a small wind tunnel (working section 0.25m × 0.25m × 2m) and was expressly intended to test the physics of flow reversal in the deep canopy so that a full PBL simulation was not necessary.

4   Scalar Flow and Transport in a Tall Canopy on a Low Hill

4.1   Analytical model

The extension of the solution of HLR to scalar flow and transport over a low hill (RWCH) treated exchange of a general scalar with a rough hill and then applied it to calculate radiant energy partition into sensible and latent heat. It assumed that the driving wind field was provided by the theory of HLR. In this section we briefly sketch an extension of the approach of RWCH, in which the rough surface is replaced by a canopy and the driving wind field is given by FB. The mean scalar concentration $\bar{c}$ and its eddy flux f are divided into background and hill-induced perturbations as follows,

$$\bar{c} = C_B(z) + \Delta \bar{c}(x,z); \quad f = F_B(z) + \Delta f(x,z) \qquad (10.24)$$

The background scalar field in the canopy $C_B(z)$ is the solution of the one-dimensional mass conservation equation,

$$0 = -\frac{\partial F_B(z)}{\partial z} + \chi(z) \qquad (10.25)$$

The scalar source term is expressed as,

$$\chi(z) = a(z)g[U_B(z)][C_0 - C_B(z)] \qquad (10.26)$$

where C_0 is a reference concentration assumed constant on the surface of the foliage, and $g[U_B(z)]$ is the leaf-level boundary layer conductance, which has a power law dependence on wind speed, $g = AU_B^n$. When the leaf boundary layer is laminar, $n \simeq 0.5$ and when it is turbulent $n \simeq 0.8$ (Finnigan and Raupach 1987). It is convenient to define a parameter γ such that $g = \gamma U_B$, where $r = 2\gamma/C_d$ is a leaf level Stanton (Nusselt) number, characterizing the relative efficiencies of scalar (heat) and momentum transfer to the leaf. In most situations $r \sim 0.1$ or smaller.

We parameterize the background scalar flux $F_B(z)$ using a mixing length model as we did for momentum,

$$F_B = -l^2 \frac{\partial U_B}{\partial z} \frac{\partial C_B}{\partial z} \qquad (10.27)$$

As was the case for momentum also, the use of a mixing length is merely a convenient heuristic in the mean flow but is physically correct when used to model the scalar perturbations caused by the hill. Using Equations 10.11, 10.26 and 10.27, the mean canopy mass conservation Equation 10.25 becomes,

$$\frac{\partial^2 C_B(z)}{\partial z^2} + \frac{\partial C_B(z)}{\partial z} = r[C_0 - C_B(z)] \qquad (10.28)$$

The solution to Equation 10.28 for small r applies within the canopy[4] and above the canopy is matched smoothly to the conventional logarithmic profile to give,

$$C_B(z) = \begin{cases} C_0 + \dfrac{c_*}{k} \ln(\dfrac{z+d}{z_c}) & \text{for } z > 0 \\[1em] C_0 + (C_h - C_0) \exp(\dfrac{r\beta z}{l}) & \text{for } z \leq 0 \end{cases} \qquad (10.29)$$

where $C_h = C_B(0)$ and the other parameters are fixed by matching the scalar profiles and fluxes at the canopy top whence,

$$F_c(0) = -u_* c_* = -r\beta^2 (C_h - C_0) U_h; \quad c_* = r\beta(C_h - C_0)$$
$$d = l/k; \quad z_c = d\exp(-\frac{k}{r\beta}) \qquad (10.30)$$

When we compare exact and numerical solutions of Equation 10.28 for $n = 0.5$, 0.8 and 1.0 we see minor quantitative but no qualitative changes in the profiles of $C_B(z)$. Similarly, for $r \sim 0.1$ there is little difference between the exact profiles and the 'small r' approximation. Exact solutions will be compared with these approximations in a forthcoming publication.

By comparing Equations 10.29 and 10.30 with the comparable expressions for the wind speed, Equations 10.11 and 10.12 we see that the leaf level Stanton number r plays a significant role in the scalar solution

---

[4]Note that Equation 10.29 applies in the limit of $n \to 1$, i. e. well mixed canopy airspace but the overall form of the solution is relatively insensitive to variation of n within the limits $1 > n > 0.5$.

and that since r is small, the background scalar concentration through the canopy varies substantially more slowly than the background wind speed.

Above the canopy we adopt the solution of RWCH for the scalar perturbations. This follows the pattern of HLR by dividing the flow field into an outer layer and a shear stress layer of depth h_i. In the outer layer, scalar perturbations are governed by inviscid dynamics while in the shear stress layer, changes to the scalar flux also play a role at first order. The linearized equations for the scalar perturbation induced by the hill are,

$$\text{Outer layer} \quad U_B \frac{\partial \Delta \bar{c}}{\partial x} + \Delta \bar{w} \frac{\partial C_B}{\partial x} = 0 \qquad (10.31)$$

which implies that $\Delta \bar{c}(x,z)$ in the outer layer is entirely the result of distortion of the isopycnals of $\bar{c}$ as streamlines converge and diverge.

In the shear stress layer the divergence of the eddy flux of $\bar{c}$ becomes important so the linearized mass balance is,

$$\text{Shear stress layer of depth } h_i \quad U_B \frac{\partial \Delta \bar{c}}{\partial x} + \Delta \bar{w} \frac{\partial C_B}{\partial x} = -\frac{\partial \Delta f}{\partial z} \qquad (10.32)$$

RWCH showed that in this region two other mechanisms become important in determining $\Delta \bar{c}(x,z)$. The first is the changes induced by the hill in the eddy flux field $\Delta f(x,z)$ and hence in its divergence. The second is the change to the flux of $\bar{c}$ from the surface that occurs because the surface shear stress $\Delta \tau(x,0)$ varies as the hill is traversed. The mechanism for this in the rough hill model of RWCH is the representation of the surface flux by a flux-gradient expression, $f(x,0) = -K \partial \bar{c}/\partial z$, involving the scalar diffusivity $K_c(x,z) = k u_*(1 + \Delta \tau/2)z$. In the shear stress layer this modulation of the surface flux boundary condition is the dominant influence on $\Delta \bar{c}(x,z)$ but when a canopy is present, this boundary condition is supplanted by the canopy dynamics.

As was the case for the momentum field, the scalar perturbation in the canopy divides asymptotically into a linearized upper-canopy solution and a non-linear, lower-canopy solution. In the upper canopy the perturbation mass balance to $O[H/L]$ becomes,

$$\text{Upper canopy layer} \quad 0 = -\frac{\partial}{\partial z}\left[l^2\left(\frac{\partial U_B}{\partial z}\frac{\partial \Delta \bar{c}}{\partial z} + \frac{\partial \Delta \bar{u}}{\partial z}\frac{\partial C_B}{\partial z}\right)\right]$$
$$-\frac{r}{2L_c}[U_B \Delta \bar{c} + \Delta \bar{u} C_B] \qquad (10.33)$$

and we see that in the upper canopy advection is small compared to the flux divergence and source terms.

In the lower canopy the flux divergence becomes small as both $U_B(z)$ and $C_B(z)$ decay exponentially. However, a sensible velocity perturbation $\Delta\bar{u}$ continues to drive the scalar source term so that the lower-canopy mass conservation equation becomes,

$$\bar{u}\frac{\partial \bar{c}}{\partial x} = \frac{r}{2L_c}|\bar{u}|\bar{c} \qquad (10.34)$$

Equation 10.34 like its momentum equivalent, Equation 10.18, is non-linear but for a different reason. At leaf level the boundary layer conductance g depends only on the magnitude of the wind velocity, $\bar{u}$ not on its direction so that we must write $g = A|\bar{u}|^n$. In the upper canopy, where $U_B > \Delta\bar{u}$ this dependence on absolute velocity need not be made explicit, vide Equation 10.34 but in the lower canopy, where $\bar{u} \simeq \Delta\bar{u}$, it is critical. We will not develop the full solutions for the scalar perturbation field here but we can describe the results of the different dynamics in the various layers qualitatively. We will take the case of a canopy sink, that is $S_0 < S_B(z)$. For a canopy source, $S_0 > S_B(z)$ the signs of the perturbations are reversed.

- *Outer layer:* In this layer the dynamics are inviscid and streamline convergence over the hill crest brings isopycnals of $\bar{c}$ from higher in the boundary layer closer to the surface. With the canopy a sink, $\bar{c}$ increases with height so above h_i we see a positive (increased) value of $\Delta\bar{c}$ over the crest in phase with the perturbation in velocity (Equation 10.20).

- *Lower canopy layer:* In the lower canopy the velocity field is dominated by $\Delta\bar{u}$ which is a maximum around $x = -L$ (upwind of the crest) and a minimum at $x = +L$ (downwind of the crest). However the sink (negative source) term depends on the absolute magnitude of $\Delta\bar{u}$ through $g = A|\bar{u}|^n$ so that the sink is large both upwind and downwind of the crest at $x = \pm L$ and small on the hill crest $x = 0$. At the same time the velocity is +ve upwind and -ve downwind of the crest, leading to flow convergence towards the hilltop. Through the advection term $\bar{u}\partial\bar{c}/\partial x$ on the LHS of Equation 10.34 this convergence combines with the maxima in the sink strength upwind and downwind to effect a minimum in $\Delta\bar{c}$ at the hill crest.

- *Upper canopy layer:* The dominant term in the upper canopy is the effect on the sink of the velocity perturbation through the last term of Equation 10.33, $r/2L_c[\Delta\bar{u}C_B]$. Since $\Delta\bar{u}$ peaks at the hill crest, this produces a minimum in $\Delta \bar{c}$ there. This effect can be interpreted as the canopy counterpart of the dominant role played by changes in surface stress on a rough hill in modulating the surface flux boundary condition, and thereby $\Delta \bar{c}$, as discussed earlier. Hence, as a result of quite different dynamics, the dominant influences on $\Delta \bar{c}$ in the lower and upper canopy layers are in phase leading to a minimum in $\Delta \bar{c}$ at the hill crest. This simple picture is modulated by other effects in the upper canopy layer, however. The eddy flux divergence couples the upper canopy to the shear stress layer above, where advection is important at first order (Equation 10.32) and within which the contributions to $\Delta \bar{c}$ that are caused by canopy dynamics must decay to match the inviscid, streamline convergence effects around $z = h_i$.

The overall analysis yields the typical magnitudes of the velocity and scalar perturbations within the canopy, U_c and C_c, respectively,

$$U_c = \frac{U_0^2 H L_c}{U_h L^2}; \qquad C_c = r U_c \frac{C_h}{U_h} \qquad (10.35)$$

Note that the magnitude of the velocity perturbation depends on the driving pressure gradient which is $O[U_0^2 H/L^2]$ and is determined by the outer layer flow as well as by the momentum absorption in the canopy, characterized by L_c and U_h. For the case we have analyzed, the scalar perturbations are caused entirely by the wind field and not by variations in the biological source/sink strength and we have ensured this by choosing a constant concentration boundary condition C_0 on the foliage surface. We see then that the scalar perturbations are relatively smaller than the velocity perturbations that drive them, the proportionality factor being the leaf-level Stanton number, r.

The foregoing analysis together with Equation 10.35 now allows us to non-dimensionalize the mass balance equations in a way that permits us to compare the expected magnitude of the advection terms at a given flux tower site with the eddy flux, $F_B + \Delta f = \overline{w'c'}$, the term we usually measure,

$$\underset{\left[\frac{U_0^2}{U_h^2} \frac{H}{L} \frac{L_c^2}{L^2}\right]}{\underset{\text{Horiz. Adv.}}{U_B \frac{\partial \Delta \bar{c}(x,z)}{\partial x}}} \underset{\left[\frac{U_0^2}{u_*^2} \frac{H^2}{L^2} \frac{L_c}{L}\right]}{\underset{\text{Vert. Adv.}}{+ \Delta \bar{w} \frac{\partial C_B}{\partial z}}} \underset{1}{\underset{\text{Eddy Flux Diverg.}}{+ \frac{\partial F_B(z)}{\partial z}}} \underset{\left[\frac{U_0^2}{u_*^2} \frac{H^2}{L^2} \frac{L_c}{L}\right]}{+ \frac{\partial \Delta f(x,z)}{\partial z}}$$

$$0.09 \qquad\qquad 0.09 \qquad\qquad 1 \qquad\qquad 0.09$$

$$\text{Scalar Source/Sink} = \chi_B(z) + \Delta\overline{\chi}(x,z) \quad (10.36)$$

$$\begin{array}{cc} 1 & \left[\dfrac{U_0^2}{U_h^2}\ \dfrac{H}{L}\ \dfrac{L_c}{L}\right] \\ 1 & 0.08 \end{array}$$

The dimensionless groups are formed by scaling the terms in the equation using L as the characteristic scale of horizontal variation so that $\partial/\partial x \sim 1/L$ and taking l as the scale of vertical variation. We see that the divergence of background eddy flux and the background canopy source/sink are both of order one as expected on a low hill. Note also that although the Stanton number r plays a significant role in determining the absolute magnitude of the individual terms in the mass balance, it disappears from their ratio. Finally we see the horizontal and vertical advection terms and the vertical divergence of the perturbation eddy flux are of the same order. The magnitude of these dimensionless groups is calculated in the bottom row of Equation 10.36 for a canopy-covered hill with the following parameter values:

$$L = 100 \text{ m}; \quad H = 10 \text{ m}; \quad L_c = 5 \text{ m}; \quad u_* = 0.5 \text{ m s}^{-1};$$
$$U_h = 1.67 \text{ m s}^{-1}; \quad U_0 = 6.85 \text{ m s}^{-1} \quad (10.37)$$

For this particular choice of canopy density, $L_c = 5$ m, the magnitude of the perturbations induced by the hill is of order of the hill slope, H/L. In a canopy with $h_c = 20$ m this corresponds to a leaf area index (LAI) of 4. If LAI = 2 but the other parameters remain unchanged, the magnitude of the perturbation terms all double because the ratio of the momentum absorption distance L_c to the hill length scale L plays an important dynamic role in determining the velocity perturbations that drive the scalar fluctuations in the canopy.

While these dimensionless groups are useful in signaling when topographically driven advection may be a problem at a given flux site, they do not tell the full story because the streamwise variation of the various terms in Equation 10.36 may lead to them canceling or reinforcing at different positions on the hill. We illustrate this in Figure 10.7, where we have plotted the streamwise variation of the terms in Equation 10.36 calculated at the canopy top with the parameter values as in Equation 10.37 above with the exception of L_c, which is set to 10 m.

In Figure 10.7a we plot the individual terms for the case of a canopy sink. We see that the horizontal and vertical advection terms are of the same order and vary roughly in anti-phase while the eddy flux reaches its maximum negative value just ahead of the hill crest, which is where the

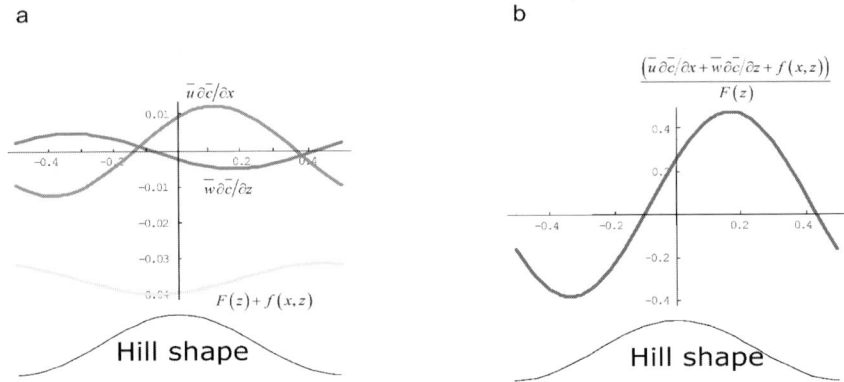

Figure 10.7. (a) Variation of the horizontal and vertical advection terms and the eddy flux over a sinusoidal hill covered with a tall canopy. (b) Fractional error entailed in estimating the canopy sink S from the background eddy flux, $F(z)$ alone.

velocity peaks as we saw in Figure 10.2. In Figure 10.7b we present these results in a different way by taking the local ratio of the eddy flux at the tower position, $x = 0$ with the sum of the three terms we would not normally measure at a flux station with a single tower, i. e. the integral over the canopy height of the first, second and third terms on the LHS of Equation 10.36. The canopy height is taken as $h_c = 20$ m but because of the exponential decay of the aerodynamic terms in the canopy, the value of h_c is not critical so long as it is $\geq 2L_c$ (see Equation 10.23). We can take this local ratio as a rough measure of the error incurred in estimating $\overline{\chi}$ in an advective situation when only the vertical eddy flux is measured. Clearly, the position of the tower relative to the hill is critical with a possible error of ±40% depending on wind direction.

In this section we have used an analytic theory to understand the advection caused by flow distortion and the corresponding changes to the eddy flux structure and source/sink strengths in the most basic case, where a generic scalar has a constant concentration on the foliage surface. To maintain an analytic solution we have made a series of somewhat draconian simplifying assumptions. Within these limitations, it seems that even relatively gentle topography can have a significant effect on the mass balance and especially on the terms used to deduce canopy-atmosphere exchange at flux towers. In the next section we will relax these simplifications by using a numerical solution to explore the effect of a realistic boundary condition for photosynthesis.

4.2 Numerical model

Deflection of the airflow over a hill perturbs the eddy flux and introduces advection terms into the mass balance. It also affects the source/sink terms by modulating the boundary layer conductance on the leaves. In the case of important scalars such as carbon dioxide and water vapor the source/sink strengths at leaf level are also subject to strong ecophysiological controls. Katul et al. (2004a) (henceforth KFL) set out to compare the magnitudes of hill-induced distortions to these ecophysiological processes with the aerodynamic processes discussed in the last section. They used a numerical approach that not only allowed a more realistic boundary condition at the leaf but also removed the approximations used in the analytic flow model.

The wind field and scalar flow and transport models were those described in Sections 3 and 4.1 above and employed the same parameterizations with the following exceptions: a) it was not assumed that r was small; b) the exponent of the boundary layer conductance was set to 0.5, i. e., $g = A\bar{u}^{0.5}$; c) the method of matched asymptotic expansions was not necessary so that advection, turbulent stress and scalar eddy flux were calculated everywhere in the flow domain. It is worth noting here that the assumptions embodied in the analytic wind field model are supported by numerical calculations as shown by Katul et al. (2004b).

In the analytic model the boundary condition in the canopy was that $\bar{c}(x,z) = C_0$ on the foliage. Instead KFL use a physiological boundary condition for carbon dioxide assimilation and respiration so that f_c, the flux of c across the leaf surface, is described by a stomatal conductance, g_s in series with g,

$$f_c = \frac{c_i - \bar{c}}{g^{-1} + g_s^{-1}} \quad (10.38)$$

where c_i is the intracellular CO_2 concentration. The stomatal conductance g_s is dependent on net leaf photosynthesis (A_n) and can be parameterized using multiple semi-empirical formulations. Based on a recent study by Katul and Chang (1999), KFL selected the Leuning (1995) model for g_s because it best describes the stomatal response to vapor pressure deficit. A_n is described using the Farquhar et al. (1980) and Collatz et al. (1991) models. The functional forms of these expressions for g_s and A_n are given in Appendix A.

Photosynthesis, A_n is driven by the photosynthetically active radiation Q_p. The vertical attenuation of incident radiation $Q_p(x,z)_{z<0}$ through the canopy is computed using the method of Campbell and Norman (1998), while expressions from RWCH were used to model the variation of $Q_p(x,0)$ across the hill as a function of the solar elevation

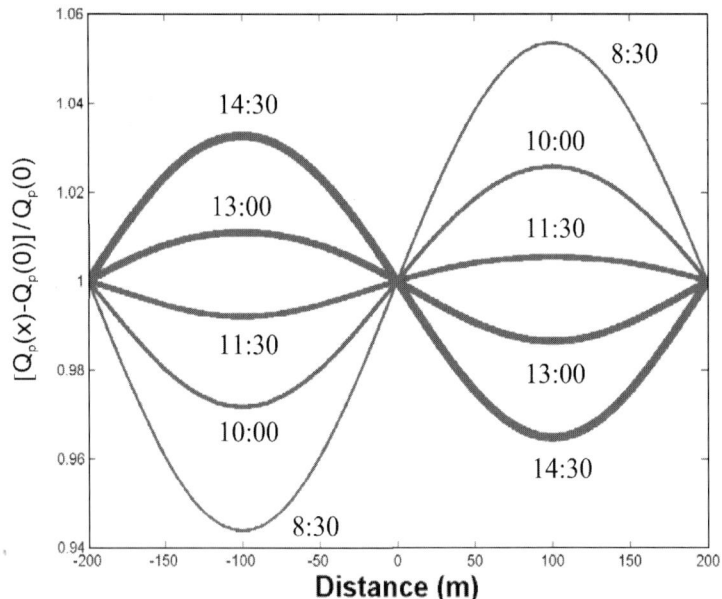

Figure 10.8. Relative variation in incident photosynthetically active radiation across the canopy top at different times of day for the north-south orientated sinusoidal hill ($H = 20$ m; $L = 100$ m). Figure reproduced from Katul et al. (2004b).

and azimuth and the local hill slope. These formulae are given in Appendix B. Finally, the soil respiration was specified. KFL compared a constant concentration boundary condition of 600 ppm CO_2 at the soil surface with a physically correct but computationally more troublesome flux boundary condition. The differences in the overall results between the two boundary conditions were minor and in the examples that follow it is the concentration boundary condition that is used.

To assess how topography alters the spatial distribution of photosynthesis and NEE, KFL applied the model described above to a gentle sinusoidal two-dimensional ridge orientated in the north-south direction ($H = 20$ m, $L = 100$ m). Next, the model calculations were repeated for flat terrain ($H = 0$) using identical radiative and canopy drag attributes and ecophysiological parameters. KFL used published ecophysiological, respiration, and drag properties obtained from studies of the Duke Forest AmeriFlux site, a mid-latitude broadleaf forest with $h_c = 10$ m and LAI = 4 (see Katul and Chang 1999, Lai et al. 2000 and 2002). The case shown below corresponded to a time of 14.30 hours, at which time

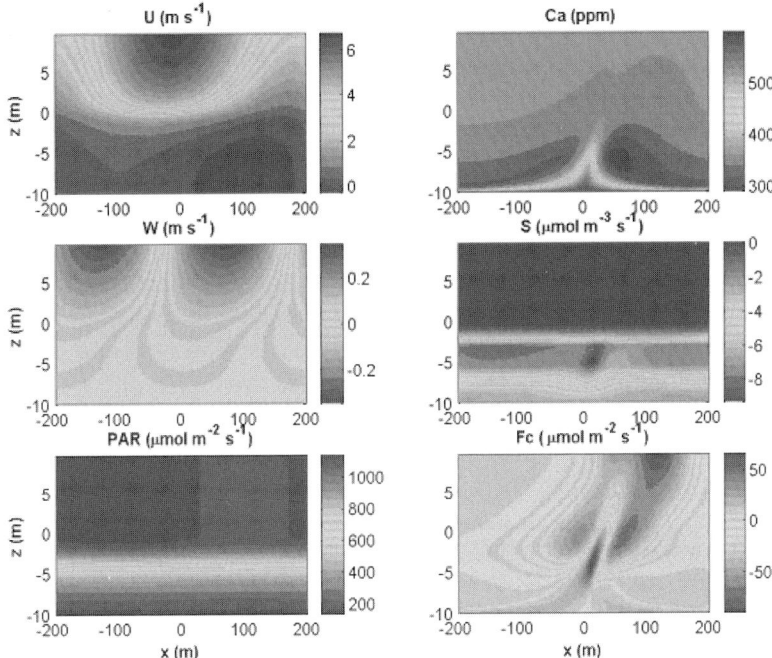

Figure 10.9. Variation of primary forcing variables, $\overline{u}(x,z)$, $\overline{w}(x,z)$, and $Q(x,z)$, where Q is the photosynthetically active radiation, PAR (left hand panels), with the response of the CO_2 concentration, C_a, the canopy photosynthesis, S and the eddy flux of CO_2, $\overline{w'c'} = F_c$ (right hand panels). Figure after Katul et al. (2004b).

the variation of $Q_p(0,x)$ across the north-south ridge was $\pm 6\%$ with a maximum on the upwind slope (Figure 10.8). Finally, u_* was set at 0.4 m s^{-1}.

The basic forcing variables and the scalar response in and above the canopy are summarized in Figure 10.9. The first panel, showing $\overline{u}$, demonstrates the speed-up to the hilltop and the marked flow asymmetry in the canopy caused by deceleration and reversed flow in the lee at ground level. Variation in $\overline{u}$ is complemented by changes in $\overline{w}$ manifested as inflow to the canopy upwind of the crest and outflow in the lee as shown in panel 2. The third panel on the left shows that the variation of $Q_p(0,x)$ over the hill is small above the canopy and indiscernible within.

The top panel on the right hand side shows the response of $\bar{c}(x,z)$. What is particularly marked is the plume of high concentration of $\bar{c}$ that is drawn up from the forest floor by the lower-canopy convergence process described in Section 4.1. This plume is diffused once it enters the upper canopy, where the eddy flux of c is active, and bent downwind by advection in the shear stress layer. Also marked is the depletion of c in the upper canopy behind the hilltop, where the recirculating flow reduces ventilation of the canopy space. The next panel shows the CO_2 sink strength, $S \equiv \chi$ and in contrast to the large changes in the $\bar{c}$ field, only minor variations of S in the x-direction are seen as photosynthesis responds to streamwise modulation of g and $\bar{c}$. The reason for the conservative behavior of S is the control exerted by $Q_p(0,x)$ together with the limits placed on photosynthesis by the electron transport capacity of the enzyme Rubisco (Appendix A). The last panel, showing the eddy flux $f(x,z)$ is the most surprising result. Here we see that despite the fact that the canopy is an overall sink for CO_2, around the plume of high concentration of $\bar{c}$ just behind the hilltop, the eddy flux $\overline{w'c'}$ is positive.

We recall from Equation 10.2 that the mass balance in a notional control volume is computed by integrating the point-valued mass balance over the volume V. On a single tower we can only perform the integration in z (Lee 1998) so in Figure 10.10 we have presented the components of the mass balance integrated from the ground to the canopy top. The first panel shows the integrated canopy sink $\int_{-h_c}^{0} S\,dz = \text{NEE}$. This of course is the term we are trying to deduce by difference from the other components of the mass balance. First we note that the canopy sink shows a significant variation as we traverse the hill, peaking at roughly 42 μmol m^{-2}s^{-1} just behind the hill crest in our example. Averaged over the entire hill, however, the NEE is roughly 35 μmol m^{-2}s^{-1}, only slightly increased over the NEE we calculate for the same forest on flat ground, which we see is roughly 32 μmol m^{-2}s^{-1}. This difference of about 10% is the same order as the hill slope, H/L, which is what the analysis in Section 4.1 would lead us to expect.

The top panel on the RHS of Figure 10.10 shows first that for the flat terrain case the eddy flux at the top of the canopy exactly balances the NEE as we expect from Equation 10.2 in the steady state case. Over the hill in contrast, the average eddy flux is substantially smaller than the NEE and we can see that this is caused by the large positive value of F_c just behind the crest, a feature we noted in Figure 10.9. Indeed, if we attempted to deduce NEE from a measurement of $\overline{w'c'}$ in the region just behind the hilltop we would get a result not just of the wrong magnitude but of the wrong sign. The reason for this is seen in

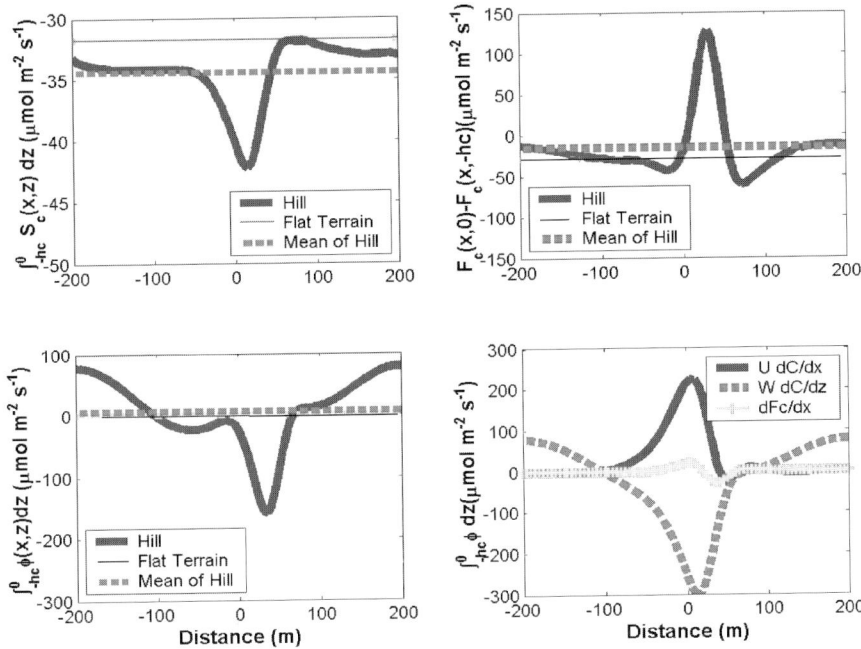

Figure 10.10. Components of the mass balance for photosynthesis after Katul et al. (2004b). Panels are a) upper left: total photosynthesis in canopy on hill compared with same canopy on flat ground: local variation (solid line); average over the hill (dark dashed line); flat ground (light dashed line). b) upper right: eddy flux from the canopy only (soil flux is subtracted). c) lower left: Sum of the advective plus horizontal eddy flux divergences. d) lower right: individual advection and horizontal eddy flux divergence terms.

the lower panel on the LHS where the sum of the advective terms and the horizontal divergence of streamwise eddy flux are plotted. These three terms have an even larger but oppositely signed peak behind the crest and the sum of these three terms (which we cannot measure on a single tower) with the eddy flux term is equal to the x-wise variation in NEE shown in the first panel. In the last panel we break the sum of the terms that vary in x into its constituent parts and see that, integrated through the canopy, $\overline{u}\partial \overline{c}/\partial x$ and $\overline{w}\partial \overline{c}/\partial z$ are roughly equal and opposite with $\int_{-h_c}^{0} (\overline{w}\partial \overline{c}/\partial z)\, dz$ negative in this case, while $\int_{-h_c}^{0} (\partial \overline{w'c'}/\partial z)\, dz$ is an order of magnitude smaller than either advective term.

In summary, the effect of the hill on the wind and radiation fields and the attendant changes in $\overline{c}$, g and Q_p lead to changes in the average NEE over the hill of order H/L and local variations about three times larger.

The effect of the wind field changes on the terms that we use to deduce NEE from a flux tower is much more severe, to the extent that, if we cannot measure the advection terms, a tower placed near the hill top and recording only $\overline{w'c'}$ could get both the sign and magnitude of NEE quite wrong. In this case NEE is the relatively small difference of two large aerodynamic terms, both of which must be accurately recorded.

5   Discussion and Conclusion

At first glance the results in the last section are alarming. However, while sounding a strong note of warning for flux measurements from single towers in complex topography, they do represent a worst-case scenario. First of all, wind perturbations generated by flow normal to a two-dimensional ridge are larger than arise in any other configuration. If we take $\Delta S_{\max} = [\Delta \overline{u}(0,z)/\overline{u}(-\infty,z)]$, the maximum speed-up of the upwind flow to the hilltop, as a measure of the wind perturbation, then on a two-dimensional ridge, $\Delta S_{\max} \simeq 2H/L$ while on an axisymmetric hill $\Delta S_{\max} \simeq 1.6H/L$ and over a two-dimensional escarpment $\Delta S_{\max} \simeq 0.8H/L$ (Kaimal and Finnigan 1996). As we saw in Sections 3 and 4, the characteristic velocity perturbation in the canopy, U_c is directly proportional to the velocity perturbation in the outer layer and the scalar perturbation C_c is proportional to U_c (Equation 10.35). Second, the combination of soil respiration of CO_2 and canopy assimilation results in maximum contrast between the 'plume' of high CO_2 concentration that originates through convergence to the crest in the lower canopy flow and the ambient $\overline{c}$ values in the upper canopy and shear stress layers. This contributes to the strong local perturbation in eddy flux, $\overline{w'c'}$ we saw near the hilltop. Hence we can expect the size of the advective terms relative to $\overline{w'c'}$ to be smaller in three-dimensional topography and for species other than CO_2.

Nevertheless, there remains room for significant error, if the advection terms are not measured at sites in hilly terrain. Falk et al. (2000) have measured the advection directly at the Wind River AmeriFlux site and found that in the prevailing wind directions it accounts for up to 40% of NEE. Furthermore, daytime assimilation is one of the most important measurements made at FLUXNET sites so that the combination of plant assimilation and soil respiration will be encountered daily.

The turbulence parameterizations used in the analytic and numerical models are mixing lengths, the simplest appropriate forms. These are convenient but physically incorrect heuristics when applied to the mean flow but can be shown to be physically correct when applied to the perturbations (FB). More accurate turbulence parameterizations have been developed for uniform canopies and a non-equilibrium second-order closure model, applicable to inhomogeneous canopy flows is also available (Ayotte et al. 1999).

All these approaches use parameterizations that implicitly assume a spectral gap between turbulence and mean flow inasmuch as the turbulence is treated as a stochastic field that interacts with the 'mean' field but is clearly distinct from it. To compare models of this kind with measurements we need to ensure that this separation applies to the measured data too. Hence, if the background fields are time varying, we must separate means and turbulent components by a filter rather than a block time average because, as Figure 10.2 shows, time averaging results in slow variations in the 'mean' fields being treated as part of the turbulence. In such a case, comparing measured and modeled turbulent fluxes would be to compare apples and oranges.

A serious limitation on the development of advection models of whatever complexity is the dearth of experimental data against which to compare them. To measure advective flux divergence in the field, an array of towers is needed. Such experiments are very expensive and so far are rare, particularly at tall canopy sites where tall towers are required. We have already mentioned the measurements at Wind River. Aubinet et al. (2003) have performed measurements at the Vierselm CarboEurope site, concentrating on nighttime gravity currents but no other multi-tower data are currently available in the open literature. Because of the expense of field experiments, wind tunnel modeling seems an attractive alternative but here too, experiments are few and confined entirely to the wind field. Ruck and Adams (1991) and Neff and Meroney (1998) have made measurements above the canopy over model hills but the only published study of the turbulent wind field within and above the canopy on a hill remains that of Finnigan and Brunet (1985). To date, no one has published measurements of scalar flow and transport in a canopy on a model hill. It is clear that until this need for data is addressed either by wind tunnel or field experiments, model development will be slow and the fidelity of the models uncertain.

Finally we point out that the models of scalar flow and transport over hills that we have presented provide a mechanism for the generation of 'low frequency' eddy flux over flat or gentle topography. Sakai et al. (2001), Finnigan et al. (2003) and Malhi et al. (Chapter 5) have shown the importance of such low frequency contributions to surface exchange. The structure of the linearized models discussed in Sections 3 and 4 is such that the deflection of the flow over the hill defines the forcing pressure field but from then on the calculations proceed as if this pressure field were applied to a canopy on flat ground. (All linearized hill-flow models have this mathematical structure). The first-order pressure field produced by a sinusoidal hill is itself sinusoidal and so any stationary pressure forcing in the boundary layer can be decomposed into a superposition of such Fourier modes and the resulting scalar and wind field perturbations calculated by the 'flow-over-hills' theories we have described. Large scale pressure fields that are convecting downwind can be handled in the same framework. Indeed, if the time scale of convection $T_{\text{conv}} (= L_p/U_{\text{conv}}$, where L_p is the characteristic length of the pressure field and U_{conv} its convection velocity) is large compared to the relaxation time of the surface layer turbulence, no modification to the theories is necessary. The theories predict the development of covariances between w and c as we have seen and these correspond to low frequency contributions to the eddy flux. The source of such large-scale pressure fields is itself worthy of investigation. Obvious candidates are tropospheric motions of various kinds and contrasts in surface energy balance as clouds pass by.

6 Appendix A: Model for Stomatal Conductance

The Leuning (1995) model for the stomatal conductance g_s is given by,

$$g_s = g_0 + \frac{a_1 A_n}{\bar{c}_s - \Gamma^*}(1 + \frac{D_s}{D_0})^{-1} \tag{10.39}$$

where Γ^* is the CO_2 compensation point, $\bar{c}_s$ is the mean surface CO_2 concentration, which can be related to $\bar{c}$ using $\bar{c}_s = \bar{c} - A_n/g_b$, D_s is the mean surface vapor pressure deficit and g_0, a_1, and D_0 are constants that vary among plant species.

The canonical forms of the Farquhar et al. (1980) and Collatz et al. (1991) models for A_n are given by,

$$A_n = \min \begin{cases} \dfrac{\alpha_p Q_p e_m(\bar{c}_i - \Gamma^*)}{\bar{c}_i + 2\Gamma^*} - R_d \\ \dfrac{V_{c\max}(c_i - \Gamma^*)}{\bar{c}_i + k_c(1 + \frac{o_i}{k_0})} - R_d \end{cases} \tag{10.40}$$

with α_p the leaf absorptivity for PAR, e_m the maximum quantum efficiency, Q_p the PAR irradiance, $V_{c\max}$ the maximum catalytic capacity of Rubisco, k_c and k_0 the Michaelis constants for CO_2 fixation and O_2 inhibition with respect to CO_2, and o_i and c_i the leaf oxygen and CO_2 concentrations, respectively. $R_d \approx 0.015 \times V_{c\max}$ is the respiration rate of the foliage. The latter constants vary with temperature as described in Lai et al. (2000) for example.

7 Appendix B: Model for Photosynthetically Active Radiation

To compute A_n, the variation in PAR, Q_p is required. The vertical attenuation of incident radiation $Q_p(0,x)$ through the canopy is given by (Campbell and Norman 1998),

$$\frac{Q_p(z,x)}{Q_p(0,x)} \approx \exp(-\Omega \times K_b(x',\Psi) \times \int_0^z a(z)\mathrm{d}z) \quad (10.41)$$

where in a first order analysis, the LHS of Equation 10.41 is the fractional amount of light arriving at depth z within the canopy, K_b is the extinction coefficient, which depends on the zenith angle (Ψ) and x', the projected leaf area, which is defined as the ratio of the projected areas of canopy elements on horizontal and vertical surfaces. Finally, Ω is the clumping factor.

For an ellipsoidal leaf distribution,

$$K_b(x',\Psi) \approx \frac{\sqrt{x'^2 + \tan^2(\Psi)}}{x' + 1.774 \times (x' + 1.182)^{-0.733}} \quad (10.42)$$

For spherical, vertical, and horizontal leaf angle distributions, $x' = 1$, 0, and ∞, respectively.

$Q_p(x,0)$ varies across the hill as a function of the solar elevation angle (Ψ_e), the azimuth angle (Ψ_a), and the local hill slope [$\alpha(x)$], and is given by RWCH,

$$\frac{D[Q_p(0)]}{Q_p(0)} = [-\cot(\Psi_e)\cos(\Psi_a)\sin(\alpha) + \cos(\alpha) - 1] \quad (10.43)$$

where $D[.]$ denotes a change in a quantity across the hill.

8 Acknowledgment

The author is indebted to his colleagues, Dr. Stephen Belcher, University of Reading, UK, Dr. Gabriel Katul, Duke University, USA, and Dr. Yves Brunet, INRA Bioclimatologie, Bordeaux, France for the use of material from our joint papers and to Mr. Dale Hughes, CSIRO Atmospheric Research for his skill in performing the wind tunnel experiments referenced in Section 3.

9 References

Aubinet, M., Heinesch, B. Yernaux, M.: 2003, 'Horizontal and vertical CO_2 advection in a sloping forest', *Bound.-Layer Meteorol.* **108**, 397-417.

Ayotte, K. W., Finnigan, J. J., Raupach, M. R.: 1999, 'A second-order closure for neutrally stratified vegetative canopy flows', *Bound.-Layer Meteorol.* **90**, 189-216.

Ayotte, K. W., Davy, R. J., Coppin, P. A.: 2001, 'A simple temporal and spatial analysis of flow in complex terrain in the context of wind energy modelling', *Bound.-Layer Meteorol.* **98**, 275-295.

Baldocchi, D. D., Falge, E., Gu, L., Olson, R., Hollinger, D., Running, S., Anthoni, P., Bernhofer, Ch., Davis, K., Evans, R., Fuentes, J., Goldstein, A., Katul, G., Law, B., Lee, X., Malhi, Y., Meyers, T., Munger, W., Oechal, W., Paw U, K. T., Pilegaard, K., Schmid, H. P., Valentini, R., Verma, S., Vesala, T., Wilson, K., Wofsy, S.: 2001, 'FLUXNET: A new tool to study the temporal and spatial variability of ecosystem-scale carbon dioxide, water vapor and energy flux densities', *Bull. Am. Meteorol. Soc.* **82**, 2415-2434.

Belcher, S. E., Jerram, N., Hunt, J. C. R.: 2002, 'Adjustment of a turbulent boundary layer to a canopy of roughness elements', *J. Fluid Mech.*, **488**, 369-398.

Bradley, E. F.: 1968, 'A micrometeorological study of velocity profiles and surface drag in the region modified by a change in surface roughness', *Q. J. Roy Meteorol. Soc.* **94**, 361-379.

Campbell, G., Norman, J.: 1998, *An Introduction to Environmental Biophysics*, Springer, 286pp.

Collatz, G. J., Ball, J. T., Grivet, C., Berry, J. A.: 1991, 'Physiological and environmental regulation of stomatal conductance, photosynthesis and transpiration: A model that includes a laminar boundary layer', *Agric. For. Meteorol.* **54**, 107-136.

Dyer, A. J., Crawford, T. V.: 1965, 'Observations of climate at a leading edge', *Q. J. Roy Meteorol. Soc.* **91**, 345-348.

Falk, M. B., Park, Y.-S., Paw U, K. T., Pyles, R. D., Hsiao, T. C., Shaw, R. H., King, T., Matista, A. A., Wahbeh, H.: 2000, 'A comparison of the carbon and water vapor exchange contributions of mean advection, eddy-covariance, and storage in a tall forest', *24th Am. Meteorol. Soc. Conf. Agric. For. Meteorol.*, American Meteorological Society, Boston, Massachusetts.

Farquhar, G. D., Von Caemmerer, S., Berry, J. A.: 1980, 'A biochemical model of photosynthetic CO_2 assimilation in leaves of C3 species', *Planta* **149**, 78-90.

Finnigan, J. J.: 2004, 'A re-evaluation of long-term flux measurement techniques. Part II: coordinate systems', *Bound.-Layer Meteorol.*, in press.

Finnigan, J. J.: 2002, 'Momentum Transfer to Complex Terrain', *Geophysical Monograph Honouring J R Philip*, (Raats, P. A. C, Smiles, D. E., Warrick, A. W. Eds) American Geophysical Union, in press.

Finnigan, J. J.: 2000, 'Turbulence in plant canopies', *Annu. Rev. Fluid Mech.* **32**, 519-571.

Finnigan, J. J., Belcher, S. E.: 2004, 'Flow over a hill covered with a plant canopy', *Q. J. Roy Meteorol. Soc.*, in press.

Finnigan, J. J., Clements, R., Malhi, Y., Leuning, R., Cleugh, H.: 2003, 'A re-evaluation of long-term flux measurement techniques. Part I: Averaging and coordinate rotation', *Bound.-Layer Meteorol.* **107**, 1-48.

Finnigan, J. J., Brunet, Y.: 1995, 'Turbulent airflow in forests on flat and hilly terrain' *Wind and Trees*, (Coutts M. P., Grace, J. Eds.) Cambridge University Press, UK, 3-40.

Finnigan, J. J., Raupach, M. R.: 1987, 'Transfer processes in plant canopies in relation to stomatal characteristics', *Stomatal Function* (Zeiger, E., Farquar, G. D., Cowan I. R.) Stanford University Press, Stanford, California, 385-429.

Hunt, J. C. R., Leibovich, S., and Richards, K. J.: 1988, 'Turbulent shear flow over low hills', *Q. J. Roy Meteorol. Soc.* **114**, 1435-1470.

Jackson, P. S., Hunt, J. C. R.: 1975, 'Turbulent wind flow over a low hill', *Q. J. Roy Meteorol. Soc.* **101**, 929-956

Katul, G. G., Chang, W. H.: 1999, 'Principal length scales in second-order closure models for canopy turbulence', *J. Appl. Meteorol.* **38**, 1631-1643.

Katul, G. G., Finnigan, J. J., Leuning, R., Belcher, S. E.: 2004a, 'The influence of hilly terrain on canopy-atmosphere carbon dioxide exchange', *Bound.-Layer Meteorol.*, in review.

Katul, G. G., Finnigan, J. J., Belcher, S. E.: 2004b, 'Momentum transfer within a canopy situated on complex topography' *Bound.-Layer Meteorol.*, in review.

Kaimal J. C., Finnigan J. J.: 1994, *Atmospheric Boundary Layer Flows: Their Structure and Management*, Oxford University Press, New York. 289 pp.

Kuo, Y. H., Schlatter, T. W.: 1990, 'Mesoscale data assimilation', *Notes from an NCAR Summer Colloquium*, NCAR, Boulder, Colorado. pp. 644

Lai, C. T, Katul, G., Butnor, J., Ellsworth, D., Oren, R.: 2002, 'Modelling nighttime ecosystem respiration by a constrained source optimization method', *Global Change Biology*, **8**, 124-141.

Lai, C. T., Katul, G. G., Oren, R., Ellsworth, D., Schäfer, K.: 2000, 'Modeling CO_2 and water vapor turbulent flux distributions within a forest canopy', *J. Geophys. Res.* **105**, 26333-26351.

Lee, X.: 1998, 'On micrometeorological observations of surface-air exchange over tall vegetation', *Agric. For. Meteorol.* **91**, 39-49.

Leuning, R.: 1995, 'A critical appraisal of a combined stomatal-photosynthesis model for C3 plants', *Plant, Cell, and Environment* **18**, 339-355.

Nastrom, G. D., Gage, K. S.: 1985, 'A climatology of atmospheric wave number spectra of wind and temperature observed by commercial aircraft', *J. Atmos. Sci.* **42**, 950-960.

Neff, D. E., Meroney, R. N.: 1998, 'Wind-tunnel modeling of hill and vegetation influence on wind-power availability', *J. Wind Eng. Industrial Aerodyn.* **74-76**, 335-343.

Raupach, M. R., Weng, W. S., Carruthers, D. J., Hunt, J. C. R.: 1992, 'Temperature and humidity fields and fluxes over low hills', *Q. J. Roy Meteorol. Soc.* **118**, 191-225.

Raupach, M. R., Finnigan, J. J., Brunet, Y.: 1996, 'Coherent eddies and turbulence in vegetation canopies: the mixing layer analogy', *Bound.-Layer Meteorol.* **78**, 351-382.

Rider, N. E., Philip, J. R., Bradley, E. F.: 1963, 'The horizontal transport of heat and moisture: a micrometeorological study', *Q. J. Roy Meteorol. Soc.* **89**, 507-531.

Ruck, B., Adams, E.: 1991, 'Fluid mechanical aspects of the pollutant transport to coniferous trees', *Bound.-Layer Meteorol.* **56**, 163-195.

Sakai, R. K., Fitzjarrald, D. R., Moore, K. E.: 2001, 'Importance of fow-frequency contributions to eddy fluxes observed over rough surfaces', *J. Appl. Meteorol.* **40**, 2178-2192.

Schimel, D. S., Kittel, T. G. F. Running, S., Monson, R., Turnipseed, A., Anderson, D: 2002, 'Carbon sequestration studied in western US mountains', *EOS* **83**, 445-449.

item Van der Hoven, I.: 1957, 'Power spectrum of horizontal wind speed in the frequency range from 0.0007 to 900 cycles per hour', *J. Meteorol.* **14**, 160-164.

Van Dyke, 1978: *Perturbation Methods in Fluid Mechanics*, Parabolic Press, Stanford, pp 271.

Index

Active sensor, 135
Advection
  estimate for carbon dioxide flux, 236
  generated by source distribution, 217
  generated by topography, 218
  generated by wind field, 218
  modeling, 211
  order of magnitude estimate, 231, 237
  relationship to averaging length, 115
  relationship to coordinate system, 43, 50, 53
  relationship to WPL, 130
Advective acceleration, 162
Aerodynamic drag, 219–221, 234
Aerodynamic method, 8
Aircraft, 106, 186, 188
AmeriFlux, 1, 72, 77, 95, 234, 238
Analytical model, 210, 218–219, 232
Atmospheric surface layer, 34, 45, 47, 70, 103, 109, 120, 126, 167, 169, 189
Attenuation factor, 137, 139, 141, 146
Autocorrelation function, 169–170
Autocovariance, 188–189
Averaging length, 12, 19–20, 68–70, 73, 75, 79, 102, 106, 109, 114–115, 128, 130, 188–189, 214–215
Averaging operator
  control volume, 35, 39, 43–44, 120, 122–123, 212, 236
  ensemble averaging, 164, 169, 214
  Reynolds averaging, 11, 69, 75, 78, 121, 164, 172–173, 187–188, 213
  spatial averaging, 124, 172, 176, 217

Bandwidth, 68
Bare soil, 167–169
Beta function, 74
Bin-averaging, 85
Binormal, 40–42
Biomass inventory, 202
Birch, 84
Body force, 162–163
Boundary condition

concentration, 230, 234
flux, 228, 230, 234
for regional scale modeling, 211
physiological, 233
sensitivity to, 162, 174–175
Broadness parameter, 73–76, 81, 83

Canopy layer, 174, 176, 221–222, 228–230
Canopy roughness, 151, 171
Capping inversion, 105, 171
CarboEurope, 1, 72, 84–86, 196, 202, 239
Carbon dioxide
  affected by gravity wave, 174
  compensation point, 240
  cospectrum, 79, 82, 85–86, 88, 90, 92
  flux adjustment, 199
  flux comparison, 56–57
  influenced by averaging schemes, 19, 24–26
  instrument gain drift, 23
  instrument zero drift, 22–23
  intracellular concentration, 233
  Michaelis constant, 240
  mixing ratio, 128
  modeled flux, 234, 236
  soil flux, 234, 238
  wavelet analysis, 112–113
Cartesian coordinate
  advantages and disadvantages, 43
  conservation equation in, 37, 40
  control volume, 122
  definition, 40
  notation, 210
Chewamegon, 20
Closed-path analyzer
  attenuation factor, 141
  conversion to mixing ratio, 127
  distinction from open-path, 135, 137
  integral correction factor, 146
  pressure fluctuation, 140
  signal processing, 142
  temperature fluctuation, 140
  transfer function, 143

245

volume averaging, 138, 145, 147
WPL correction for, 127, 140
Closure model, 239
Closure problem, 48
Cloud, 174
Cloud street, 105
Clumping factor, 241
Coherent structure, 109, 174, 199, 201
Complex terrain
  concentration field in, 236
  contribution to low frequency, 117
  cospectrum in, 81
  eddy flux field in, 231, 236
  fairly thin shear layer, 38
  GLEES, 77
  Great Mountain, 55
  Griffin, 84
  in analytical model, 219, 226
  in numerical model, 233
  in wind tunnel, 210, 224, 226
  PAR in, 235, 241
  relationship to averaging and filtering, 27
  velocity field in, 222, 224–225, 235
  Vierselm, 239
  Weidenbrunnen/Waldstein, 197
Conductance
  boundary layer, 226, 229, 233
  model for, 233, 240
  stomatal, 233, 240
Confidence level, 169
Conservation equation
  at a point, 36–37, 121, 217
  for moist air, 120
  in Cartesian coordinate, 37, 40, 217
  in homogeneous flow, 9, 37, 126, 217
  in streamline coordinate, 42
  integrated over a control volume, 123, 213
  linearized, 228
  order of magnitude estimate for, 230
  scale analysis of, 230
  vertical integration of, 9, 124, 236
  volume integration of, 122
Continuity equation, 122, 217
Control volume, 35, 39, 43–44, 120, 122–123, 212, 236
Convective boundary layer, 186, 188
  height, 105–106, 109
  velocity scale, 105
Coordinate line, 39–41, 43–44
Coordinate system
  Cartesian, 37, 40, 122, 210
  closure problem, 48
  effect on low frequency contribution, 116
  flux comparison among, 56–57
  for point measurement, 44
  global and local properties, 34, 44, 52
  guiding requirements, 36

  instrument, 45
  loss of information, 48
  natural wind, 47
  over-rotation, 48
  planar fit, 48
  relationship to averaging operation, 35, 116, 213
  relationship to data quality, 48, 199
  relationship to filtering, 28, 116
  streamline, 40–41
  surface following, 43
  vector basis, 34, 39–40, 42–43, 52, 62
Coordinate tilt
  examples of, 53, 115
Coriolis parameter, 192
Correlation coefficient, 164, 190, 193
Cospectral similarity, 21, 70–72, 84
Cospectrum model, 73
Cospectrum
  in complex terrain, 81
Covariance
  equation for, 12, 189
Cross-wind effect, 46
Curvilinear, 39, 42

Data assimilation, 211–212
Deardorff's relationship, 105
Decoupling parameter, 110
Density covariance, 136
Detrending, 12, 187–188, 213
Dirac delta, 9, 36, 212
Directional derivative, 42
Douglas fir, 84
Downdraft, 105–106
Drag coefficient, 219
Dry air flux, 122–124
Duke forest, 174, 234
Dynamic temperature, 192

EBEX, 200
Eddy shedding frequency, 149
EdiRe, 85
Edisol, 85
Effect-level ring, 196
Energy balance
  equation, 198
  in LES, 108
  lack of, 29, 114, 116, 153, 198
Engelmann spruce, 77
Ensemble average
  for coordinate rotation, 48, 115
Ensemble averaging, 164, 169, 214
Entrainment, 103, 171
Enzyme Rubisco, 236, 240
Equation of state, 126

INDEX

Ergodic hypothesis, 166
Ergodicity
  definition, 166
  in weak sense, 166
  necessary conditions for, 169, 171
Exponential profile, 219
Extrapolation, 93

Fairly thin shear layer, 38
Fetch, 84, 193, 203
FFT, 79, 83, 85
Filter
  antialiasing, 142
  Bessel, 142
  block averaging, 15, 69, 93
  boxcar, 116
  definition, 14
  FIR, 27
  linear detrend, 15
  moving average, 15, 213
  recursive, 16, 71, 75
First order instrument, 142
Flow convergence, 54, 229
Flow interference, 45, 47, 190, 194
Flow model
  analytical, 210, 219, 232
  numerical, 233
  physical, 212
Flow reversal, 224–225, 235
Flux-variance relationship, 172, 190–192
Flux bias
  caused by advection, 53, 218, 231–232
  caused by averaging schemes, 11, 19, 26, 114
  caused by coordinate tilt, 11, 50–52, 56, 115
  caused by density effects, 119, 150, 153
  caused by gap filling, 11, 200
  caused by spectral attenuation, 11, 68, 187
  in LES, 108
  nighttime, 52, 71, 130, 164, 199
  relationship to averaging length, 116
Flux chamber, 200, 202
FLUXNET-Canada, 1
FLUXNET, 1, xiii, 182, 195, 202–203, 210, 218, 238
Foliage area, 219, 241
Footprint, 182, 184, 190, 196–197
Fourier transform, 14, 18
Free atmosphere, 103
Free convection, 48, 51, 103
Frenet Frame, 40
Friction velocity, 3, 51–52, 164, 192, 199, 219
Frost, 187

Gain function, 142–144
Gap filling, 11, 51, 200–201, 203
Gaussian, 185
Geostrophic wind, 107
Glacier Lakes Ecosystem Experiment Site (GLEES), 77
Grassland, 194
Gravity wave, 174–175, 177, 185, 188, 190, 201
Great Mountain, 55
Grid cell, 196
Griffin, 23, 26, 84–85, 95, 112–113

Haar transform, 111, 186
Hamming window, 85
Heat storage, 198
Horizontal eddy flux, 45, 52, 236
Hypothetical experiment, 164

Ideal gas law, 126
Inertial subrange, 73, 80–81, 167
Initial condition, 161–163
Instrument coordinate
  cross-wind effect in, 46
  vector basis for, 46
Instrument malfunction, 47, 185, 187, 199, 201, 203
Instrument offset, 22–23, 50, 54, 62, 186
Integral turbulence characteristics, 189–193
Interaction layer, 103
Intermittent turbulence, 185, 188
Internal boundary layer, 130, 186, 190, 217
Interpolation, 201, 210
Inviscid flow, 220, 228–230
Isotropy, 167, 177

Jaru, 11, 112–113

Kaimal spectrum, 21, 70, 75, 81–83, 85
Katabatic forcing, 110
Kurtosis, 79, 186

Large eddy simulation, 107–108, 116
Laser-doppler anemometer, 225
Latent heat flux, 194, 197–198
Leaf area index (LAI), 84, 231
Leaf boundary layer, 226
Least squares, 14, 50, 54, 62, 80, 85
Leonard term, 17, 213
LIDAR, 167, 170–171
Lindenberg, 195
Line averaging, 68, 71, 88, 128–130

LITFASS, 194
Local acceleration, 162
Locally scalable, 172
Logarithmic profile, 219, 227

Maize, 49
Manaus, 11, 19, 24–25, 114
Matching layer, 103
Meandering motion, 188
Mechanical mixing, 138
Mesoscale motion, 45, 113, 187
Microfront, 185
Micrometeorology
  definition, xiii
  flat-earth paradigm, xiii, 210
Mixing layer, 201
Mixing layer analogy, 222
Mixing length, 219, 222, 227, 238
Mixing ratio, 121
Molar concentration, 120, 128
Molecular diffusion, 36, 121
Moment, 165–166, 169
Momentum absorption distance, 219, 231
Momentum flux
  affected by coordinate rotation, 56–57
  affected by gravity wave, 174
  cospectrum, 81, 84
  cross-wind, 47, 55, 57
  divergence, 222
Monin-Obukhov length, 51, 78, 103, 192
Monin-Obukhov similarity, 52, 109, 169
Morning transition, 22
Moss, 84
Moving average, 15, 213
Moving equilibrium hypothesis, 172
Multiple regression, 187

Natural wind coordinate
  definition, 47
  example of large rotation angles, 57, 199
  rotation procedure, 60
  underlying assumption, 34, 47
Navier-Stokes equation
  in poor man's form, 162
Net radiometer, 198
NIGEC, 2
Nighttime flux
  correction based on friction velocity, 51, 164, 199
Non-stationarity, 171, 175–177, 184, 190, 201
Nonlinear regression, 74, 80
Nonlinearity, 163
Normalization parameter, 73
Numerical model, 233
Nusselt number, 226

Ogive test, 20–21, 93, 189
Omnidirectional, 55, 193
One-dimensional flow, 34, 36, 39, 50, 52, 125, 212, 217
Open-path analyzer
  distinction from closed-path, 135, 137
  WPL correction for, 129, 138
Orthonormal, 40, 42
Outer layer, 103, 220, 228
Outer time scale, 109
Over-rotation, 48, 54
OzFlux, 1, 28

Passive sensor, 135
Pathlength averaging, 46, 187
Perturbation
  pressure, 219
  scalar concentration, 227
  shear stress, 219
  wind, 219
Phase function, 142, 144
Photosynthesis, 109, 233, 235–236
Photosynthetically active radiation, 233, 241
Physical model, 212
Pine, 174
Planar fit coordinate
  definition, 48
  example of, 49, 57
  Matlab code, 63
  rotation angles, 49
  vector projection scheme, 62
Point-by-point density correction, 128, 154, 157
Polygonal pattern, 105
Potential flow, 220
Potential temperature, 103
Power law, 73–74
Precipitation, 187, 201
Pressure fluctuation
  dynamically induced, 149
  in detection chamber, 131, 147
  in sampling tube, 148
  static, 149–150
Pressure
  cospectrum, 82
  covariance, 78–79, 126, 136
  gradient, 162, 221–222, 225
  perturbation, 219
  sensor, 78
Principal normal, 40

Quality flag, 193, 195, 197

INDEX

Random error, 11, 108, 117, 174, 187–188
Random number generator, 87–88
Randomness, 162
Realization, 162, 164, 166, 169
Recursive filter, 16, 71, 75
Resonance cavity, 148
Reynolds averaging, 11, 69, 75, 78, 121, 164, 172–173, 187–188, 213
  reasons for, 214
  rules, 213
Reynolds number, 85, 146, 148
Rossby wave, 214
Rotation angle, 52, 55, 57, 115, 199
Roughness length, 196, 219
Roughness sublayer, 103, 110
Running mean, 15–16, 26

Sensible heat flux
  affected by gravity wave, 174
  correction for water vapor effect, 129
  cospectrum, 81–82
  ogive test on, 20, 93
Sensor geometry, 46
Sensor separation, 68, 71–72, 88, 128–129
Shear stress layer, 220–221, 228
Signal to noise ratio, 167
Site characterization, 183
Sitka spruce, 84
Skewness, 79, 185
Slope parameter, 73, 80–81
Soil heat flux, 198
Sonic anemometer
  ATI, 78
  CSAT3, 194
  Gill/Solent, 84, 186
  Kaijo Denki, 55
Source term, 9, 36, 120, 217
Source weight function, 196
Spatial averaging, 124, 172, 176, 217
Spectral gap, 102, 214–215, 239
Spike, 11, 79, 184–185, 203
Standard deviation
  of temperature, 172, 192
  of wind components, 52, 192
Stanton number, 226
Stationarity
  definition, 164, 166, 190
  in stable conditions, 171
  in weak sense, 166
  lack of, 20, 22, 25
  relationship to homogeneity, 166
  test of, 169, 184, 187, 190–191
Streamline, 38
Streamline coordinate
  advantages and disadvantages, 43
  conservation equation in, 42

defintion, 40
Strouhal number, 149
Student t-test, 169
Subalpine fir, 77
Sublimation, 77
Surface layer, 34, 45, 47, 70, 103, 109, 120, 126, 167, 169, 189
Surface renewal, 201
Surface roughness, 106

Tall tower, 20
Tapering, 80
Taylor's hypothesis, 111
TEAL structure, 108–109
Temporal averaging, 69, 75, 78, 121, 164, 172–173, 187–188, 213
Tethered balloon, 106
Thought experiment, 139
Time constant, 15–18, 25–26, 71, 75, 89, 140, 142–143, 145, 148
Time lag, 11, 137
Transfer function
  for block averaging, 15–16, 69
  for closed-path analyzer detection chamber, 142
  for linear detrend, 15–16
  for mean removal, 69
  for running mean, 15–16
  method to correct high frequency loss, 69
  pseudo, 17
Triple moment, 121
Tube flow, 11, 90, 148–149
Tumbarumba, 28
Turbulent organized structure, 106, 117

Updraft, 105–106

Vapor pressure deficit, 233, 240
Vertical velocity
  due to mesoscale motion, 45
  in LES, 106
  in WPL, 126
  over hills, 232
Vierselm, 239
Virtual temperature, 79, 103, 129
Viscosity, 163
Volume averaging, 124, 217
Von Karman constant, 103, 219
Vortex shedding, 149

Water budget, 202
Water vapor
  affected by gravity wave, 175

concentration, 128
correction for density effect, 129
cospectrum, 79, 82, 86
Lidar measurement of, 167
mixing ratio, 128
quality flag, 197
stability correction function for, 170
Wavelet transform, 111
Weidenbrunnen/Waldstein, 196–197
Weighting factor, 196
Well-developed turbulence, 193
Well-mixed conditions, 51
Wind directional shear, 45, 47–48, 224–225, 235
Wind River, 46, 238–239
Wind tunnel, 162, 210, 212, 224, 226
Window function, 14

Windowed correlation coefficient, 164
Windowing, 80
WPL
  correction to carbon dioxide flux, 127–128, 130
  correction to water vapor flux, 127–129
  for closed-path analyzer, 127, 140
  for open-path analyzer, 129, 138
  in one-dimensional form, 127
  influenced by pressure fluctuations, 126, 147
  point-by-point correction, 128, 154, 157
  relationship to advection, 124, 130
  relationship to bias in time integrated flux, 150, 153
  relationship to spectral attenuation, 133

Zero-plane displacement, 11, 77, 130, 219